Center for the Study of National Reconnaissance Classics

A History of Satellite Reconnaissance
The Perry Gambit & Hexagon Histories

CENTER FOR THE STUDY OF
NATIONAL RECONNAISSANCE
CHANTILLY, VA

JANUARY 2012

Preface

Coinciding with the commemoration of the 50th Anniversary of the National Reconnaissance Office (NRO), the Director of the NRO, Mr. Bruce A. Carlson, publically announced the declassification of the Gambit and Hexagon imagery satellite systems on 17 September 2011. This announcement constituted the NRO's single largest declassification effort in its history. The Gambit and Hexagon programs were active for nearly half of the organization's history by the time of the declassification announcement. Their history very much represents the NRO's history—one that is defined by supremely talented individuals seeking state of the art space technology to address difficult intelligence challenges.

The United States developed the Gambit and Hexagon programs to improve the nation's means for peering over the iron curtain that separated western democracies from east European and Asian communist countries. The inability to gain insight into vast "denied areas" required exceptional systems to understand threats posed by our adversaries. Corona was the first imagery satellite system to help see into those areas. It could cover large areas and allow the United States and trusted allies to identify targets of concern. Gambit would join Corona in 1963 by providing significantly improved resolution for understanding details of those targets. Corona provided search capability and Gambit provided surveillance capability, or the ability to monitor the finer details of the targets.

For many technologies that prove to be successful, success breeds a demand for more success. Once consumers of intelligence—analysts and policymakers alike—were exposed to Corona and Gambit imagery, they demanded more and better imagery. Consequently, the Air Force, who operated the Gambit system under the auspices of the NRO, entertained proposals for an improved Gambit system shortly after initial Gambit operations commenced. They received a proposal from Gambit's optical system developer, Eastman Kodak, for three additional generations of the Gambit system. Ultimately the Air Force settled on only developing the proposed third generation because the proposed second generation offered minimal incremental improvement and the fourth generation appeared technologically unachievable at the time. The third generation became known as Gambit-3 or Gambit-cubed while it was under development. Once it replaced the first generation, it simply became Gambit. The new Gambit system, with its KH-8 camera system, provided the United States outstanding imagery resolution and capability for verifying strategic arms agreements with the Soviet Union.

Corona was expected to serve the nation for approximately two years before being replaced by more sophisticated systems under development in the Air Force's Samos program. It turned out that Corona served the nation for 12 years before being replaced by Hexagon. Hexagon began as a Central Intelligence Agency, (CIA) program with the first concepts proposed in 1964. The CIA's basic goal was to develop an imagery system with Corona-like ability to image wide swaths of the earth, but with resolution equivalent to Gambit. Such a system would afford the United States even greater advantages monitoring the arms race that had developed with the nation's adversaries. The system that became Hexagon faced three major challenges. The first was development of the technology, which was eventually overcome by the Itek and Perkin-Elmer Corporations. The second was bureaucratic, deciding how the CIA and Air Force would cooperate in building such a system because they each had strengths and weaknesses in the development of national reconnaissance systems. The third challenge was to secure the resources that were required to build the most complicated and largest reconnaissance satellites at the time. By 1971 the NRO overcame the challenges to successfully launch the Hexagon satellite and fulfill, or even exceed, expectations for unparalleled insight into capabilities of United States adversaries.

At the time of the Gambit and Hexagon declassification announcement, the NRO released a number of redacted documents and histories on its website to the general public on the stunning Gambit and Hexagon systems. Two of the histories are contained in this volume. They are histories of the Gambit and Hexagon systems written by Air Force historian, Robert L. Perry.

Perry wrote five volumes of history related to the National Reconnaissance Office. They include volumes on the NRO's involvement in the Samos and Corona Programs as well as histories of early national reconnaissance efforts. The former Office of History of the NRO released a redacted and edited version of Perry's history of early NRO management. The history provided unique insight into early management of national reconnaissance systems just prior to the formation of the NRO and in the organization's early years.

When Perry prepared individual volumes on the Gambit and Hexagon programs, they constituted the third volume of his series. The volume was broken into Part A for Gambit and Part B for Hexagon. The histories were released in redacted forms when details about the KH-7, KH-8, and KH-9 programs were released several years ago. The versions in this publication have been reviewed based on updated guidance in conjunction with the programmatic declassification of the Gambit and Hexagon systems.

For many years, Perry's histories have proven to be a significant resource for not only those concerned about the history of national reconnaissance, but for the practitioners of national reconnaissance as well. Perry's work captured the difficulties associated with the development of early national reconnaissance systems, and more importantly, the strategies for meeting those challenges. A reader of a Perry history cannot help but come away with a better understanding of national reconnaissance programs, their management, and how to be more effective in their own job.

Perry's histories also serve as exemplars of the art and craft of historians. They are rich in detail, well-sourced, and written with engaging prose. This is true for the Gambit and Hexagon histories. Perry devotes considerable effort to telling the stories of each program based on the documentary evidence available to him at the time of the writing and those he was able to speak to about the programs. As a consequence, Perry's readers come away with insight into the technological opportunities presented by pursing both programs, the organizational tension that arose between the Air Force and CIA because of competing approaches for imagery satellites, and the force that human determination is for dealing with unknowns and set-backs that are hallmarks of efforts to acquire large, sophisticated technological systems.

We must include a couple of editorial notes to the reader. First, with respect to redacted material, we have edited the volumes to smooth the flow of language in the volume, rather than indicate where material was redacted. In some cases, that meant removing entire sentences. Readers of this volume can find the unedited, redacted versions at NRO.gov if they are interested in reviewing where information was redacted. Second, we have preserved Perry's references to other volumes in his national reconnaissance history series. Redacted versions of all volumes are available at NRO.gov. We hope over time to complete a similar effort to this for the remaining Perry histories.

Perry completed his Gambit and Hexagon histories in the early 1970's. The systems would continue to serve the nation until the middle of the 1980's. Although these histories do not cover the entire length of Gambit and Hexagon service, they cover a significantly important part—the development and initial operation of the systems. The systems became reliable means for addressing difficult intelligence challenges once they became operational. They provided the nation reliable vigilance from above until the next generation of imagery satellites advanced the United States intelligence collection capabilities.

James D. Outzen, Ph.D.

Chief, Historical Documentation and Research
The Center for the Study of National Reconnaissance

CENTER FOR THE STUDY OF NATIONAL RECONNAISSANCE

The Center for the Study of National Reconnaissance (CSNR) is an independent National Reconnaissance Office (NRO) research body reporting to the NRO Deputy Director, Business Plans and Operations. Its primary objective is to ensure that the NRO leadership has the analytic framework and historical context to make effective policy and programmatic decisions. The CSNR accomplishes its mission by promoting the study, dialogue, and understanding of the discipline, practice, and history of national reconnaissance. The CSNR studies the past, analyzes the present, and searches for lessons-learned.

Table of Contents

A History of Satellite Reconnaissance: The Perry Gambit History

- Gambit Preface 10
- Gambit Origins and Development 12
    - Endnotes 36
- Gambit - 1 Operations 39
    - Endnotes 58
- The Development and Operation of Gambit-3
    - Background and Nomenclature 60
    - The Origins of Gambit-3 60
    - Development Flights 69
    - Policy, Administration and Further Development 75
    - Flight Program-Vehicles 7 to 22 79
    - Flight History of Gambit-3 Block II (Double-Bucket) 83
    - Endnotes 88

A History of Satellite Reconnaissance: The Perry Hexagon History

- Hexagon Preface 96
- Introduction and Background 98
- Evolution of a System 101
- Hexagon: Program Onset to first Flight 108
- Hexagon: Initial Operations 121
- Hexagon: Program Reorganization and Product Improvement (1971 - 1973) 128
    - Endnotes 132

Center for the Study of National Reconnaissance Classics

A History of Satellite Reconnaissance
The Perry Gambit History

Robert L. Perry
November 1973

Gambit Preface

Gambit was conceived while Dwight David Eisenhower was President of the United States. Thirteen years later, when this preface was written, the system still was the principal reliance of the United States government for surveillance of areas to which that country was denied access. It was, of course, a vastly different system from that first proposed shortly after Gary Powers' U-2 ran afoul of a Soviet antiaircraft missile in May 1960. At the time of that incident, the United States had no operational reconnaissance satellites and of the two developmental systems with apparent neartime potential, Samos E-1 was conceptually flawed and the other, Corona, had experienced a frustrating succession of operational failures. Four additional photo-satellites (Samos E-2, E-3, E-4, and E-5) were at some stage between invention and first launch; none was ever to return a single photograph of Soviet territory to American photo interpreters, although that preposterous outcome could not then have been foreseen by any rational participant.

U-2 penetrations had provided some useful insights into the research and development status of Soviet missile and aircraft programs by 1960, but the United States desperately needed information about the characteristics, numbers, and placement of operational ballistic missiles in the Soviet inventory. Notwithstanding the urgency of that need, President Eisenhower chose to disapprove plans for further U-2 operations over Russia rather than chance a nuclear weapons confrontation. In any case, the vulnerability of the U-2 was all too apparent. Lacking credible information about Soviet capabilities, the United States had in 1958 undertaken an enormous expansion and acceleration of its own ballistic missile program, hopeful that American industry could overcome what was generally assumed to be a substantial Soviet advantage in nuclear weapons delivery capability. No Corona satellite had yet functioned correctly; in mid-1960 that program was forced to retreat from launching operationally configured payloads to a resumption of engineering test flights, sans cameras, in the hope that malignant defects in orbital and recovery functions might be identified and eliminated.

In the near panic that followed the discovery that U-2 aircraft could no longer safely overfly the Soviet Union, intelligence specialists devised three major new photo-reconnaissance programs: Oxcart (the Mach 3, 100,000-foot-altitude aircraft that became better known as the A-11 "Blackbird" and later fathered the SR-71 and F-12 programs), Samos E-6 (designed originally to replace the languishing Corona satellite), and Gambit. Political constraints finally kept Oxcart from fulfilling its considerable promise and Samos E-6 was technically deficient, like its five Samos predecessors. Stubborn CIA and Air Force program managers working with Itek, Lockheed, and General Electric engineers rescued Corona and by late 1960 had collected the evidence needed to demonstrate that Soviet missile rattling was mostly hollow bluster. But in the end it was Gambit that brought back the information needed to proportion the Soviet-American nuclear missile balance—though that event did not become reality until three years after the crisis that fostered the program. And notwithstanding the periodic appearance of programs and proposals for programs to supplement or supplant Gambit, that system grew and prospered so mightily that 10 years after its first flight it still was the principal reliance of United States surveillance effort.

This volume contains the history of the Gambit program from conception in 1960 to tenth anniversary of first flight, in 1973. Like other volumes in this series, it is designed to stand alone in being fully comprehensible without reference to other sources, but because the several discrete elements of the National Reconnaissance Program are inextricably interrelated, the reader may find it advisable to consult one or another of those volumes for detailed information about events that impacted on Gambit without being integrals of the program.

This history was prepared under terms of a contract between the Director, Special Projects, National Reconnaissance Office (Director, Program A), and Technology Service Corporation, of Santa Monica, California. The principal author, Robert Perry, began research and wrote draft histories of the early years of Gambit while employed first by the United States Air Force and later by The Rand Corporation. He undertook revision and expansion of those sections and the addition of Gambit-3 and Gambit flight histories in 1972, in association with Robert A. Butler, a consultant to Technology Service Corporation. At various times, parts of the manuscript have been reviewed by members of the staff of the National Reconnaissance Office and of Program A. The reviewers and suppliers of both data and documents are so numerous that it is not practical to list them here. Most are mentioned in source notes following the individual chapters. To acknowledge their invaluable assistance in this way is plainly an inadequate response, but there is no feasible alternative. In any case, for such errors and oversights as may have survived the scrutiny of contributors and reviewers, the author is entirely responsible.

GAMBIT
ORIGINS AND DEVELOPMENT

Like much of the National Reconnaissance Program, Gambit was the product of technical and political ferment and international tensions that peaked during the Spring and Summer of 1960.* The need for new sources of high resolution reconnaissance photography had become critical in the aftermath of the U-2 affair and with the enforced suspension of U-2 operations over the Soviet Union. Generally, policy-making officials in the Department of the Air Force and the Department of Defense had become thoroughly disenchanted with what they had seen of the existing Samos program. Continued emphasis on "concurrency" as a program mode and a stubborn Air Force emphasis on readout rather than recovery techniques severely prejudiced the Air Force case, since both approaches were unacceptable to most officials above the level of the Air Staff. The pressures of international politics had made it quite difficult for the Eisenhower administration to openly sponsor a new or accelerated satellite reconnaissance development. Finally, attractive proposals for new orbital reconnaissance systems had appeared during the summer of 1960. Adding body to the mixture were the facts that until mid-August the Corona had not returned any photographs whatever, while the only other capsule-recovery system then under development, Samos E-5, was regarded with something less than undiluted enthusiasm by much of the technical community.

In March 1960, Eastman Kodak (EK) had privately submitted to the CIA and separately to the Reconnaissance Laboratory at Wright Field a proposal to develop a 77-inch (focal length) camera for satellite reconnaissance. In June the company proposed a 36-inch camera system to provide convergent stereo coverage of Soviet territory. EK called the latter system "Blanket."

A month later, on 20 July, Eastman submitted a modified proposal which essentially integrated the 77-inch camera with the stereo features and film recovery techniques embodied in "Blanket." That variant was called "Sunset Strip."** Dr. E. H. Land, one of the key industry authorities in the reconnaissance program, personally brought the Eastman proposal to the attention of Air Force Undersecretary J. V. Charyk, who was rapidly becoming the dominant figure in the Pentagon struggle for control of the Air Force satellite reconnaissance effort. Charyk opened direct contact with Eastman Kodak shortly thereafter. He was particularly interested in the Eastman approach because it embodied two major elements toward which he was favorably predisposed: a film-only recovery scheme, like Corona, and a very high-acuity, long focal-length camera.

In the meantime, reconnaissance specialists of The Rand Corporation had renewed their efforts to induce the Ballistic Missile Division (BMD), immediate sponsor of the Samos program, to develop a spin-stabilized reconnaissance system along the lines of a 1957 Rand proposal. In response to a request from BMD, Rand in June 1960 began working with Space Technology Laboratories (STL) on a plan to develop a system which by taking maximum advantage of available technology could be made operational in the near term. BMD interest stemmed largely from Charyk's earlier sponsorship of such an approach.

On 7 July 1960, a group of Rand and STL specialists quietly assembled at the invitation of Colonel Paul Worthman of BMD, the sub-rosa Air Force manager of the Corona activity, to discuss details of a newly conceived variant of the original spin-stabilized satellite. Rand had concluded that it would be perfectly feasible to orbit a reconnaissance satellite in the guise of a standard ballistic missile reentry body. Its real function would be hidden under the explanation that the orbiting body was being used in tests of a ballistic missile warning system. Rand's recommendation to STL covered a 1500-pound satellite carrying a 36-inch (focal length) camera system using spin stabilization to provide panoramic coverage at a ground resolution of about 17 feet. If the satellite were oriented so as to have its lens pointing directly downward while over latitude 55 North it would provide useful coverage of all of the northern hemisphere lying between 40 and 70 degrees.[1]

By early August 1960, STL had shaped the earlier scheme into a semi-formal proposal. It differed from the earlier scheme in being based on a camera with a 24-inch focal length and in certain other minor details. Apart from re-introduction of the spin stabilization mode after a lapse of two years, its chief attraction lay in the premise of operations that could be conducted most circumspectly— even though there was a degree of unreality in the notion that a pretense of warhead detection tests long could be maintained for a vehicle which remained in a relatively stable orbit instead of

---

* The resume that follows is largely an encapsulation of Chapter VI of Volume IIA. For that reason, source citations have been used only when new material was employed.

** A then-popular television series was titled "77 Sunset Strip."

reentering steeply, as an actual warhead would do.* By 25 August 1960, when the President approved the establishment of a tightly controlled secretariat-level satellite reconnaissance organization, there were three leading candidates for sponsorship as "new" systems. The E-6 program—based on "Blanket" concepts—had proceeded toward source selection while the question of who would control the total program, and at what level, was being resolved. "Sunset Strip" was being given very serious consideration—but most privately. Unlike E-6, which was rather widely known because of the source selection proceedings, "Sunset Strip" had unfolded very quietly. Nevertheless, the general structure of Eastman's proposal was treated as normal, highly classified information at Wright Field, at BMD, and in various Pentagon offices concerned with satellite reconnaissance. The third possibility, "Study 7," as STL's proposal was called, had been handled as near-covert.

President Eisenhower approved the concept of a clandestine reconnaissance satellite development in the course of the National Security Council meeting of 25 August. Theoretically, it would have been possible to select any one of the three possible systems for covert development, even E-6, since by that time Corona experience had twice demonstrated that a widely known proposal could be officially "terminated" while actually being covertly continued. But because a great many people knew of the pressure that had been applied to bring on the development of a new reconnaissance system during the summer of 1960, it seemed unrealistic to assume that everybody who was witting of earlier activity would uncritically accept an announcement that no new system was being developed. Thus a covert E-6 program was not a real possibility.

On 20 September 1960, very shortly after the Secretary of the Air Force Samos Project Office (SAFSP) had legally come into being at BMD, Charyk met with Brigadier General Robert E. Greer (the program's new military director), Colonel Paul E. Heran (chairman of the E-6 source selection board), and Lieutenant Colonel James Seay (Greer's procurement advisor). After considering all the options, they agreed that the best course was to continue both E-6 and "Sunset Strip," which had been funded at a relatively low cost "study" level for the past several weeks. Undersecretary Charyk, with the specific approval of the President, decided the 77-inch system should be developed covertly. How that should be done remained to be determined.** For the moment, the only major action was to finance "Sunset Strip" through the balance of the year.

"Study 7," renamed Bolero, was briefly continued, but in early November Charyk had ruled against any immediate development of a spin-stabilized reconnaissance satellite. By that time the notion that a "Sunset Strip" program could be concealed under the veneer of E-6 activity had evolved; the covert effort had acquired the code name Gambit.

The general premise was that E-6 and Gambit would be made to resemble one another in outward details and that the same set of contractors would be used in both programs. The E-6 source selection board thus became the shell within which Gambit contractors were chosen. Since Eastman was the originator of the Gambit camera concept and for a variety of very practical reasons had to be picked to carry on that effort, E-6 camera development was assigned to Eastman Kodak. (That would probably have been the outcome of E-6 source selection in any case.) General Electric, the most experienced firm in the area of reentry capsules, won the Gambit award and the assignment to develop a capsule for E-6. One of the early problems encountered by program managers arose from the need for specifying a satellite vehicle which would house both Gambit and E-6 equipment. The Gambit camera was considerably larger, which meant that the E-6 had to be housed in a low-density structure that offended the engineering sense of E-6 vehicle designers.

Concealment of Gambit intentions proved less difficult than had initially been feared. General Greer arranged to have Eastman's "Sunset Strip" study contract terminated with the explanation that because of the E-6 decision no further development of a 77-inch camera was required. Simultaneously Greer's office drew up a "black" agreement which authorized EK to continue the "Sunset Strip" work as a covert program—the Gambit camera development. General Electric, which like Eastman had previous experience

---

* With allowances for minor differences, there were only three combinations of basic elements which could result in a useful panoramic satellite reconnaissance system. A three-axis-stable vehicle with a panning lens and fixed-position film was one; its chief practitioner was Corona. Use of a stable vehicle in conjunction with a moving mirror or lens arrangement and film movement synchronized with lens or mirror movement was a second; the E-6 approach of late 1960 was representative. The third technique was spin stabilization, based on a fixed position lens that rotated around the longitudinal axis of the spinning satellite and relied on moving film for both panoramic effect and image motion compensation. Although spin stabilization had first been conceived in combination with a technique of film recovery in 1956 and had been the principal ingredient of The Rand Corporation's "family of recoverable reconnaissance satellites" proposed in August-November 1957, no such system had ever been developed. Spin stabilization had briefly been the favored approach to what became Corona and the short-lived "Program IIA." A variant of spin stabilization, but using television techniques rather than film, was the basis of Program 35 (Program 417), the cloud-cover surveillance satellite which began development in 1961.

** See Chapter V (Volumn I of the Perry History's) for a considerably more detailed account of the considerations affecting the covert program decision.

in "black" programs, emulated the camera contractor in concentrating all Gambit-related activity in a secure facility. Indeed, for the first weeks of Gambit activity the matter of screening Corona efforts from General Electric's Gambit people, and Gambit from Corona, was of as great concern as the concealment of Gambit from most of the Air Force.

General Electric used the story of an alternate reentry vehicle for E-6 as its initial cover. Eastman Kodak relied on a "proprietary information" screen—a device of particular effectiveness because Eastman was notorious for ferociously guarding its new industrial developments. Aerospace Corporation, which was to perform a limited systems engineering-technical direction function in both Gambit and E-6, concealed its Gambit activity under rigid need-to-know rules. Such controls were also imposed on the two main centers of Air Force activity, the program office complex in Los Angeles and Charyk's special staff in the Pentagon. There was some early difficulty in choking off the rather casual circulation of Gambit knowledge within the headquarters of the Air Research and Development Command and in the Air Materiel Command structure, leading to a couple of episodes of knuckle rapping, but by early 1961 an effective "black" environment had been built around the program.

Perhaps more important to the surprising success of the cover effort, a great deal of effort was then being expended in developing the recovery systems for E-5 and E-6 in secure but "white" settings. To the outside world it seemed obvious that the new Samos organization would be intent on the E-5 and E-6 programs, so any puzzling activity at one of the contractor plants was attributed to these interests rather than to anything new. Inside General Greer's organization, where relatively few people initially knew of Gambit, normal human preoccupation with the tasks of the moment proved a highly successful insulator against random curiosity. Most of the Air Force shared the uncritical assumption that "the establishment" could not accommodate effective internal secrecy and that because procurement and contracting had always been open matters—and "security" a special sort of club to which most cleared Air Force personnel were admitted without qualification—no large-scale development effort could possibly be concealed.

The use of E-6 as a cover for Gambit had certain disadvantages that were recognized early. In December 1960 General Greer began to worry the question of how to conceal Gambit if E-6 were cancelled. Termination of E-6 on the grounds of technical inadequacy or budget pressure was not likely for the moment, but there seemed a real possibility that the politically vulnerable "overt" satellite reconnaissance effort might be wiped out in the aftermath of an agreement with the Soviets. In that event it would be impossible to launch a Gambit payload under the pretense that it was an E-6. For that matter, such an elementary task as construction of Gambit reentry vehicles at General Electric would be difficult to conceal once the excuse of an E-6 alternate disappeared.*

Even while the program was first taking shape, then, General Greer had recognized that eventually he would have to invent cover other than E-6— some other ostensible payload and some other apparent mission—to hide Gambit. The problem continued to trouble him for several months. Among all the space programs being conducted by NASA and the Air Force, only those contained within the reconnaissance effort were significantly concealed. Routine security screened several of the "military satellites," but experience had demonstrated that for a reconnaissance program "routine security" was not enough. The apparent susceptibility of any acknowledged satellite reconnaissance program to cancellation on political grounds was particularly acute in 1960-1961. A solution more permanently satisfactory than that of pretending to be an E-6 had to be found for Gambit.[2]

In the event, only until the early months of 1961, was it feasible to pretend that Gambit activity was part of the E-6 program. The external configurations of the two remained very much alike for a time, most of the subsystems planned for E-6 were more or less adaptable to Gambit, and the on-orbit performance requirements of the two were similar. But by March of 1961 it was clear that the E-6 design was becoming fixed and that Gambit, still in gestation, was taking a different form. Maneuver capability greater than that of E-6 was added to Gambit, propulsion arrangements (for on-orbit operations) changed, and considerably less of the E-6 development became applicable to Gambit. By June of 1961, continuing evolution of the Gambit satellite had caused it to lose most of its outward resemblance to E-6, the internal arrangements were almost totally different, and relatively few of the E-6 components remained applicable.

In such circumstances the academic concern General Greer had voiced six months earlier became a real problem. Not only was there a marked technical dissimilarity between the two systems, but the possibility of a politically motivated cancellation of both E-5 and E-6 seemed greater. The desirable solution,

---

* It may be argued that the CIA had done all those things in Corona without arousing suspicion, but in fact Corona was tightly concealed under "Discoverer" for its first four years, and in any case CIA expenditures were not matters of public record, as was the case for all Air Force Samos costs in 1960.

suggested in Greer's notes of December 1960, was total disassociation from the original Samos program. Since it was not at all feasible to hide Gambit under a scientific satellite label (the suggestion horrified the CIA, ultra-sensitive to anything that might invite close scrutiny of Discoverer and thus threaten compromise of Corona), and none of the other in-progress space programs of either the Air Force or NASA afforded proper camouflage, there was no easy or obvious option.

While mulling over the contradictions between needs and possibilities, General Greer conceived an approach based in part on his earlier analysis of the problem of covert procurement. In November 1960 he had begun "black" contracting under the philosophy that since "everybody" knew it was impossible for the Air Force to buy anything expensive without going through established review and approval channels, one might do quite a lot of unsuspected buying and contracting by merely obtaining a direct authorization. It occurred to him that the solution to the Gambit quandary might be found in the same thesis. He thereby invented the concept of the "null program," a development with no known origin and no specified goal. If such a program were conducted under the aegis of a highly classified payload, it should be entirely possible to purchase boosters, upper stages, and launch services through normal channels. Because "everybody" knew that the entire reconnaissance satellite program was in Greer's keeping, the assignment of "null program" responsibility to the regular Space Systems Division (SSD) organization would serve to convince most observers that it had to have some objective other than reconnaissance. Vague references to precise land recovery (a real but secondary objective of Gambit at the time) might serve to induce suspicion that the "null program" actually had a "bombs in orbit" goal.* Putting such a cover into effect required devious scheming and a high degree of ingenuity, but by June 1961 the plan had been reduced to specifics and generally approved by Undersecretary Charyk.[3]

In July the first moves toward establishing an activity called "Program 307" were taken. Through the Air Staff, SSD received authorization to buy four "NASA type" Agena B's for launches starting in January 1963—the Agenas to be assigned to no particular space program "for the present." In August, Charyk sent a memorandum to the Air Force Chief of Staff which emphasized the need to protect the USAF's "capability to do future space projects" and which affirmed the desirability of ordering six Atlas boosters (configured to accept Agena B's) to be used starting in February 1963. Again there occurred the phrase about "not assigned to a particular space project."[4]

Apparently the matter seemed so mundane to the Air Staff that the authorizing teletype managed to get lost somewhere in the Pentagon-AFSC headquarters maze. Nearly two weeks were needed to straighten out the resulting confusion and even then it proved necessary to apply considerable pressure before organizational inertia could be overcome.[5]

Having gotten a small batch of Atlas and Agena vehicles on order, SAFSP moved to the next business—formal creation of a "null project." On 25 September 1961, the Air Force vice chief of staff directed General B. A. Schriever, AFSC commander, to establish "Project Exemplar."** That code phrase, which was classified confidential, was defined as covering four launches from the Pacific Missile Range, starting in February 1963. The authorizing message noted that the Secretary of the Air Force had separately ordered the necessary Agenas and Atlases "on an unassigned basis." "They are hereby assigned Exemplar," the teletype read.[6]

In a further exchange of teletypes, all written well in advance in General Greer's complex, the special projects office established the fact that "Exemplar" had a goal that was classified top secret. Plans to include specific references to the procurement authorizations for the Atlas and Agena purchases had to be put off because nobody in the Pentagon could locate the relevant documents; they had been lost in the course of an Air Staff reorganization during the summer of 1961. The "white" correspondence also stated requirements for the usual sort of elaborate documentation-development plans, cost projections, and the like-that had become customary for new programs. The absence of such paraphernalia would presumably have alarmed the "normal" procurement establishment, though such "documentation" was completely redundant to the SAFSP procedures.[7]

By mid-November the basic plan had been very largely put into effect, only a few of the loose ends remaining. Gambit now consisted of a succession of elements, some covert, some within the normal military classification system. All of the "white" elements were gathered under "Exemplar"—which for reasons of administrative convenience had the additional and unclassified nickname "Cue Ball."***

---

* The author ran across the formal documentation on the "null program" (then called Program 206) early in 1962, several months before being exposed to the real workings of SAFSP. Even though he was firmly convinced that parking nuclear weapons in orbit was a most irrational project, he concluded that the Air Force was actually proceeding along that line. Any other explanation of the obvious facts was, as General Greer had cannily anticipated, too illogical to deserve serious thought. (R.P. —July 1964)

** The original code word was "Quicksilver," changed because it had been used previously.
*** "Cue Ball" was chosen to add spice to conjectures about orbiting bombs and means of returning them to precise sites on the earth.

Colonel Q. A. Riepe, who had been associated with the satellite reconnaissance program at intervals since 1953, was named the "Cue Ball" program director. His assumption of technical responsibility for Gambit from Colonel Paul Heran (E-6 program director and original custodian of the Gambit development) was not complete until February 1962, however.

A complex network of nominal and actual reporting channels linked "Cue Ball" to General Greer. (Greer then had an additional duty assignment as Vice Commander of the Space Systems Division, although his primary responsibility was for the reconnaissance mission.) All contacts between "Cue Ball" and the SAFSP structure were to remain "black," as were all Gambit budget and programming matters. "Cue Ball" was the cover for booster procurement, launch services, and certain other non-sensitive aspects of Gambit that could be handled through normal military channels, thus providing a means of deceiving those channels about the true purpose of "Cue Ball." Nowhere was there an explicit description of the "Cue Ball" payload or mission. Persistent inquirers who had some plausible justification for more information were told that "another organization outside SSD is responsible for the payload." The Atomic Energy Commission was the obvious suspect since NASA, the only other candidate, was notoriously antagonistic to classified work.* The intention of the deception was to create a vague impression that the payload was either a bomb or something related to manned space flight.

"Cue Ball" was organized along the lines of a conventional SSD program, although such "normal" channels and reporting lines were for cover purposes only; actual relations with higher authority would go ("in the black") to and through Greer's SAFSP office. It was particularly important, as General Greer emphasized frequently in the early stages of setting up "Cue Ball," that personnel prominently associated with the reconnaissance effort not be seen with "Cue Ball" personnel and that the "Cue Ball" people avoid any contaminating association with satellite reconnaissance. Not all "Cue Ball" assignees were cognizant of Gambit, so internal office security was another problem.**[8]

Misdirection continued successfully through Charyk's approval of the "Cue Ball" development plan and his formal authorization of initial funding. Key individuals at various stations in the headquarters USAF and AFSC structures had been alerted to the scheme and were presumably prepared to see that the various budget, priority, and precedence authentications emerged promptly and satisfactorily. But in the slow weeks after Charyk's directive appeared, some of the carefully laid plan began to flake away. Initially, all had gone well. Charyk's directive came out on 24 November and within three days the Air Staff had taken the actions necessary for the official start of program activity. On 13 December, however, a message from the Pentagon to AFSC specified approval of an initial program, and when a non-briefed officer went to the budget branch to clear up the "misunderstanding" he was told that there were no dollars available for "System 483A"—the nomenclature assigned to "Cue Ball" for processing purposes. In the meantime, of course, Gambit had begun to run through what remained of its money. A succession of quick telephone calls patched together an interim solution, while the larger question of how to get affairs in proper order was being worked out in more detail.[9]

The original confusion was not entirely cleared up until February 1962—by which time the complications of working with an involved "classical" structure in the Air Staff had further diffused the outlines of the original "Cue Ball" plan. Though the end result was probably all to the good, it did not seem so at the time.

As part of the cover plan, General Greer had decided to have "Cue Ball" broken up into two elements, Program A and Program B. "A" would include the first four Atlas Agena vehicles and "B" the remaining six needed for the approved 10-launch effort. Through AFSC channels, General Greer conjured up a memorandum justifying a total 10-vehicle program on the assumption of one success in each of three test configurations. Such a justification went to Undersecretary Charyk, in the open. Having apparently become somewhat foggy on the precise details of the cover scheme, Charyk questioned the need. A quick briefing straightened out the misunderstanding. Separately, Greer induced SSD to ask if SAFSP would be interested in supporting work on precise land recovery, and Greer replied with a more accurate measure of his interest, thus providing for the dollars lacking in the original December 1961 authorization. By 1 February, therefore, all again seemed to be well.[10]

Unfortunately, optimism was premature. Experience of the past two months had demonstrated that it was extremely difficult to transfer program money from the line item called "advanced research and components" (where it appeared in the Air Force budget when actually intended for Gambit purposes)

---

* It does not seem to have occurred to anyone that the CIA might have been nominated.

** An odd difficulty popped up shortly after "Cue Ball" was created. Some of the reconnaissance program personnel who had become accustomed to treating all aspects of Gambit as thoroughly "black" insisted on so handling "Cue Ball"—a tendency which endangered the security of the whole elaborate scheme.

to the "support system 483A" line item. Then, within the Air Staff there still were some objections to a 10-vehicle launch program supporting something called "Cue Ball," and about which little was known, when projects which seemed to have a much more valid requirement were "being underfunded." Additionally, and unhappily, in a sycophantic flush of enthusiasm for a program which appeared to have so much secretariat support, the Systems and Logistics element in the Air Staff had put the "Cue Ball" Pentagon project office in a very conspicuous organizational spot where it could not avoid attracting unwanted attention. One or a combination of these circumstances would surely focus more light on "Cue Ball" than was desirable. The reports and briefings required of a "normal" system were troublesome enough, but if people became interested in "Cue Ball," and set about "straightening out" the program, a lot of rather vulnerable explanations might become necessary. Alternatively, and equally undesirable, many more nonparticipants would have to be briefed on the real purpose of "Cue Ball, " thus violating the basic precept of Gambit security.

Faced with this situation, Undersecretary Charyk directed that all money for SAFSP use, including the "Cue Ball" fund, be carried under a 698AJ line item entry. (Vela Hotel, the nuclear-test detection system program, was the only non-reconnaissance program so funded.) The number identified "advanced system development." Programming entries for "Cue Ball," originally listed under 483A ("supporting system"), were changed to 698AL. The E-6 program was carried as 698AN, but also funded as part of 698AJ. Whatever else might have been achieved, the nomenclature at least had become incomprehensible.

Charyk's decision was not wholly popular, partly because the association of 698AL with General Greer's projects could be rather easily established. But since the original scheme for subduing Pentagon interest in "Cue Ball" had become unworkable, it seemed as good an approach as any.

At least one officer on Charyk's staff felt that "the limited view of the SP [SAFSP] security types as to the requirements for cover and deception..." was also a factor in a security situation which he characterized as "just a couple of steps from disaster." But somehow disaster did not follow, even though no changes of any consequence were made in either views or procedures.[11]

In retrospect, the involved convolutions of Gambit deception and occasional alarms that the "cover" was disappearing seemed overdone. General Greer's original premise, that nobody would suspect the existence of a "null program" because the very concept was inconceivable to any normal staff activity, proved sound. There was an additional contributor to the success of the various moves that put "Cue Ball" in the same funds and program categories as other "Greer programs" without bringing on a security disaster: those on the inside of Gambit tended to seek complete normality as an avenue to inconspicuousness without appreciating that the regular Air Force establishment had been conditioned to accept uncritically any decision handed down, no matter how irrational. Rationality was not inherent in development decisions, nor logic a necessary ingredient of programming. Most of the Air Staff saw nothing peculiar in an arrangement that put Vela Hotel, the "Samos" program (as it was still called), and a supposed "bombs in orbit" development in a common category called "advanced systems." Nor did the average staff officer wonder whether something might be hidden under such an arrangement. As Greer had reasoned more than 15 months earlier, since there was no precedent for what he had done, it would generally go unsuspected.*

Gambit had by early March 1962 made a partial transition from a covert program to a special-security program. The distinction was by no means clear, nor was it widely appreciated, but the key to the matter was that Gambit did not pretend to be anything else, and that such pretense represented the boundary between covert and highly secure developments. It was true that Gambit inhabited a covert atmosphere, and that procurement techniques and manufacturing practices invented for covert programs continued to be used, but in reality Gambit was a highly classified program without a publicly specified payload. There were two layers of security between the 206 ("Cue Ball") project office and the real payload; peeling away the first uncovered nothing more than the second, since the "tight security payload" story was entirely true. What mattered was that a very special security category insulated the true program goal—high resolution satellite reconnaissance.

The real covert reconnaissance program, Corona, still maintained an elaborate facade of scientific inquiry even though it was becoming constantly more difficult to devise excuses for not producing at least some

---

* The expedient of combining "Cue Ball" with other "advanced system developments" apparently was proposed by Major David Bradburn of General Greer's staff. He wanted to include a few other miscellaneous items to further mislead speculation; Agena was one he suggested. The "comptroller" of the "black" programs, also wanted to add some ARPA and NASA programs. The final decision apparently was a compromise to avoid calling too much attention to the reprogramming actions involved, though to some on Charyk's staff the affair seemed to take place in a spotlight on an empty stage.[12] (Ten years later, before a standing-room audience, Major Bradburn, then a brigadier general, became the fifth man to occupy the post established for Greer.)

scientific information from a "scientific satellite program" now in its second year of successful recoveries.*

All the while the management and security structures of the Gambit program were evolving toward the "Cue Ball"-Program 206 arrangement, meaningful development continued. Between November 1960 and January 1961, the formal contracts with Eastman went from draft to signature. There were no real obstacles, although for a time Undersecretary Charyk balked at Eastman's demand for a seven-percent profit on what was essentially a time and materials contract. Greer felt that the fee was not excessive, basing his judgment both on the precedent of U-2 experience and on Eastman's "unique capability," and after he secured the concurrence of other high officials (including the Assistant Secretary of the Air Force for Financial Management and the Air Force General Counsel), the contract was so arranged.**

By mid-1961, the concept of Gambit development, and its technical details, had been worked out in detail. Essentially, Gambit differed from E-6 (to which it still maintained a technical likeness) in having substantially higher ground resolution, in possessing a capability for photographing specific targets which were off the immediate orbital track, and in being intended for land recovery. The land recovery approach, which had been an integral of the original presentation to President Eisenhower in August 1960, was intended to overcome what were considered increasingly hazardous aspects of sea recovery. There was a great deal of concern in 1961 for the possibility that a Soviet ship or submarine might reach a floating Corona capsule before rescue forces arrived or that a capsule might descend, intact, into non-friendly territory. Recovery of such a capsule might well precipitate a grave international crisis—while failure to regain possession might be the excuse for a new U-2 incident, but an echelon or two higher on the scale of risk.***

Because of its need for higher resolution, Gambit would fly somewhat lower than E-6. A photographic altitude of 90 miles was generally considered desirable. The resolution requirement imposed a need for accurate orbit maintenance over a period of several days, for more precise altitude control than in E-6, and for an ability to rotate the camera section about the vehicle's roll axis. Land recovery implied extremely precise deboost velocities and reentry programming.

The attitude control system, then in a status of advanced design, was a two-axis gimballed platform on which were mounted infrared horizon scanners and an integrating gyroscope. The horizon sensors measured pitch and roll error; the gyro measured yaw error. Control movements were dependent on several jet-nozzle apertures, with a blend of nitrogen tetroxide and hydrazine fuel providing the impulse. The system was originally designed to permit as many as 600 roll reorientations during each mission.

A set of four rocket engines, each capable of producing 50 pounds of thrust, would provide for orbit maintenance. Six more such rockets were located in the aft section of the reentry vehicle. After reorientation of the satellite by 180 degrees and a 60-degree pitchdown had been completed, the reentry vehicle would be separated from the vehicle midsection and the engines fired. A velocity meter signaled shutdown.[14]

On 1 August 1961, at about the time the shift from an E-6 cover to a "null program" was beginning and several months after the E-6 had been committed to fabrication, Eastman completed the basic design of the Gambit payload system—the camera, cassette, and associated flight instruments. Design of the orbital vehicle was very nearly complete. Launch in January 1963 still seemed feasible. Two months later, after considerable thought and a succession of detailed studies, Undersecretary Charyk approved the use of the Wendover AFB area (Utah) for land recovery operations. He also introduced two complicating requirements: provisions for both north-to-south and south-to-north photography (north-to-south was the conventional approach), and for rapid launch—on four days or less notice. The quick-launch capability was not considered essential for early shots, however.[15]

The "Cue Ball" program office had been formally established on 10 October 1961 with an initial complement of 18 officers and 8 clerks. By that time, six Atlas boosters and four Agenas were on order and arrangements to obtain four more of each were pending. The various elements of normal development program documentation either had been completed or were well along toward completion. Arrangements for special facilities were being made: pad, launch complex, and assembly buildings modifications at Vandenberg had been scheduled and the State Department had opened negotiations for an additional up-range station, needed both for controlling the orbital vehicle and for safeguarding the proposed land recovery process.[16]

---

* A rather significant reorientation in security thinking was in progress during 1962. It led, in time, to the practice of obscuring all military space flight goals by confining released information to a terse announcement of the identity of the booster and upper stage. Under such scant camouflage Corona functioned effectively and unnoticed for another 10 years.

** Probably the two key factors in the fee decision were the original National Security Council directive of 25 August, which ordered that Samos "take" be processed by "the same agency that processed U-2 take"—Eastman Kodak, and the complete absence of alternatives. Moreover, as Eastman pointed out, the firm was currently drawing a fee elsewhere. [13]

*** A Corona capsule did survive an unplanned reentry, in Venezuela, several years later—and nobody noticed.

Nevertheless, Gambit was not in an entirely happy situation. In January 1962, the Samos E-5 program was finally cancelled after a succession of launch, on-orbit, and recovery system failures. Corona was in one of its periodic spasms of operational difficulty, and the proposal for a Lanyard development was receiving generally friendly attention. (Lanyard was a re-engineered, single-camera E-5 system in Corona vehicles.) The need for Gambit—or for a system with comparable on-orbit photographic potential—was even more marked than had been the case a year earlier. Various proposals for somehow accelerating Gambit development were being considered at precisely the time when the design weight of the Gambit system had overtaken the payload lofting potential of the Atlas-Agena. (In order to reduce weight, the six forward-firing rockets of the orbital vehicle were deleted in January 1962. This compromise had the effect of restricting orbit adjust maneuvers to those based on velocity increases—going from a lower to a higher orbit.) Perhaps more significant, Gambit development was now far enough along to begin suffering the consequences of earlier errors and oversights. Vehicle stability was rapidly becoming a critical item in early 1962, the first major technical difficulty to cause real concern.[17]

Undersecretary Charyk, who was under constant pressure to get quick and effective results in the satellite reconnaissance program, wanted both to accelerate the pace of Gambit development and to improve its product. He spent 9 February 1962 in Los Angeles discussing those needs with Greer and Riepe. They concluded that program acceleration was impractical unless a considerable degradation in photography was also acceptable. Moreover, it was then becoming clear that problems of mission planning and on-orbit control would be more difficult than originally anticipated—and substantially more complex than anything previously attempted in satellite reconnaissance. Gambit would differ from all predecessors in being committed to a computer-designed operational procedure, since the precision requirements were deemed too great to be satisfied by the sort of target designation and on-orbit procedures employed in Corona and planned for E-6.[18]

Further, the infrared horizon sensors which General Electric was developing were causing particular concern. GE's preferred single-scan sensor seemed incapable of providing the required accuracy within the time limits of the development program. Rather than being able to accelerate Gambit availability, GE had to caution that delayed vehicle delivery was probable. A program slip of about four months seemed inevitable, moving the prospective first launch date from February 1963 to May of that year.

General Electric's proposal for correcting the defect was to go to a "body bound" stabilization system, essentially abandoning—at least for early flights—the concept of a stable-platform sensor separate from a camera that rolled into appropriate aiming positions. Aerospace Corporation estimated that such an expedient would double the smear potential of the system, degrading its resolution quite markedly. Although Charyk asked for a precise evaluation of the resulting degradation, he privately told Greer to find an alternative solution, preferably one involving development of a different (back-up) sensor.[19]

General Greer had suggested that option on 28 February, shortly after the initial disclosure of General Electric's development problems. He also, but somewhat reluctantly, endorsed and forwarded Colonel Riepe's proposal for an expanded test program, one involving more qualification tests, the construction of more spares for the engineering development program, the inclusion of complete hot-firing tests in the schedule, the provision of a back-up development for major elements of the command system, funding an expansion of GE's industrial facilities, and the addition of three reentry vehicles to the basic test program. All those changes seemed certain to further delay the availability of the first Gambit.

Similar but far less sweeping recommendations affecting Eastman Kodak's program went forward simultaneously. EK, though having trouble, was in less difficulty than General Electric. The chief camera program change which Greer sponsored involved the development of additional manufacturing processes for lightweight optics.

To the greater question of whether an attempt should be made to maintain schedules at the expense of system degradation, Greer provided a blunt answer: by going to an Agena-derived stabilization system it would be possible to provide for a vehicle with limited accuracy and system flexibility that would meet the February 1963 launch date. But Greer opposed such an option unless the schedule was a sole and overriding consideration, since the development had no future and any resulting photographs would be degraded. Rather than make such a compromise, the general favored accepting a four-month slip in first launch.[20]

With reluctance equal to Greer's, Charyk accepted the prospect of further schedule slippage. On 19 March 1962, he directed that a limited-accuracy dual-scan infrared system being developed by General Electric remain the primary reference for the first three flights but that a more sophisticated dual-scan system be used thereafter. He also approved starting work on a backup single-scan system and cancelled GE's study of body-bound earth-reference techniques. He accepted Greer's recommendation that nothing immediately be done about adapting the Agena-oriented Corona stabilization technique to Gambit.[21]

The options thus adopted encouraged some optimism about meeting schedules and performance requirements should the primary development systems encounter further difficulty. There was general agreement that the earliest possible date for initial launch would be May rather than February 1963.[22]

Decisions on these matters had to be made and put into effect by mid-March; Charyk was under orders to report to the President on the status and prospects of Gambit at that time. The undersecretary began his 19 March 1962 report by recalling that the objective of Gambit, "to produce satellite photography having a ground resolution of from two to three feet," was being given "an overriding priority."

He noted that the performance requirements of the system pushed the state of the satellite arts in three specific areas: lightweight optics, vehicle stability, and the complexity of orbital operations. For practical purposes, the limits of optical resolution were decided by the surface quality of the primary mirror and the scanning mirror. Gambit mirrors were larger, made to closer tolerances, and lighter than in any previous system. Thermal gradients between the reflecting surfaces and the rear supporting surfaces had forced consideration of metal rather than glass backing, further complicating the problem.

The performance of the Gambit camera depended as much on vehicle stability as on any inherent photographic quality. Pointing had to be extremely precise, requiring extreme accuracy in the horizon sensors and a stable platform gyro system that would allow the sensors to stay locked on the horizon while the vehicle rolled to point toward targets on either side of the orbital track. Because the ground swath width of Gambit cameras was only 10 miles, more photographs would be taken from a canted than from a vertical position.

The complexity of orbital operations derived from the inability of the launch system to put the orbital vehicle on a predetermined orbit with the precision required by the narrow swath width. Command programming had to be changeable in flight, and further complexity derived from the need to set highly accurate roll positions for photography on either side of the vehicle's track.

Charyk's report was relatively optimistic, although he refrained from any predictions of complete success in meeting either schedules or resolution requirements. He forecast a first flight date in May 1963 and operational readiness by August of that year. And he concluded by summarizing the measures taken to insure the readiness of Gambit at the earliest possible date,* appending the note that additional funds would be needed to see the "insurance program" through.[23]

Certain other measures were taken during March 1962 to improve prospects of program success. At Greer's insistence, General Electric reorganized its Gambit management to provide more meaningful attention from high corporate executives and to improve laboratory, assembly, manufacturing, and test procedures. Concurrently, the general put Space Technology Laboratories under contract to solve the orbital operations problem. STL would receive funds for computer work on orbit selection, mission profiles, and operational analysis. Charyk also approved these actions in March.[24]

Separately, the West Coast group arranged with Eastman Kodak to begin a backup program supporting General Electric's infrared sensor development. Unlike most other work of such nature in support of Gambit, this particular EK effort became a "white" subprogram. (The justification was a need for improved horizon sensors to support space programs generally. Equipment tests were to be conducted in the Discoverer program.)

One key reason for making the scanner work at EK "white" was the need for close correlation of Eastman and GE developments. The feasibility of such contacts was enhanced by the fact that General Electric was moving all its Gambit work to a special facility at Valley Forge, Pennsylvania. (Corona and Lanyard subsystems were to remain at GE's Chestnut Street shops, in Philadelphia.)

The facility problem, in conjunction with an alarming increase in General Electric's cost estimates, caused a minor crisis in Gambit program affairs in April and May 1962. In part, the reluctance of the Department of Defense to finance an expansion of GE facilities, including those needed for Gambit production, arose from an expectation that space and equipment released from the cancelled Advent communication satellite program could be diverted to other uses, including Gambit.[25] General Greer felt that nothing of value to Gambit would emerge from the Advent termination in time to be useful. Charyk had

---

* These were essentially the actions approved in his separate 19 March directive to General Greer and were based on Greer's 28 February recommendations.

another opinion, and a fair amount of argument was necessary to change his mind. Eventually, on 31 May 1962, the necessary formal approval was received and the facility construction at Valley Forge was able to continue, but for about a month before approval was received the question—and the soundness of Gambit delivery schedules—seemed very much in doubt.[26]

Another aspect of the apparent reluctance to commit additional money to General Electric was the steady increase in estimated program costs. Between April 1961 and April 1962, GE's estimates had gradually climbed. The contractor explained away some of the puzzling increases as arising from unanticipated technical difficulties but also conceded to "just some bad estimating." Neither Greer nor Charyk was particularly happy about a contract performance which the general charitably described as "...somewhat less than expected."[27]

Equally important to the trend of Gambit development were technical questions which had persistently bothered General Greer through the early months of 1962. They stemmed, in the main, from Greer's long-held conviction that the need for land recovery of film capsules had been considerably overemphasized. The original Gambit program directives had specified land recovery as one of the prime development objectives, for reasons which had seemed more than sufficient to the high administration officials who had conceived and pushed the notion. In the climate of 1960, when Corona recovery had been infrequent, uncertain, and expensive, land recovery seemed a useful option. But land recovery was more complicated than air catch in several ways, and the very grave risks inherent in such an approach had continued to trouble Greer. On several occasions during the first 18 months of Gambit development he had raised the issue in discussions with Dr. Charyk. In each instance, however, Charyk had acknowledged the question and reconfirmed the requirement. All of those discussions were informal. Only once did the question of an alternative to the original land recovery scheme receive consideration at the level of the program office. In January 1962 a member of the Aerospace Corporation's Program 206 contingent advised Colonel Riepe that air retrieval was "being considered," but that it was quite impractical.[28] The very considerable weight of the 206 recovery capsule as it then existed exceeded the air-catch capacity of recovery aircraft.

By July of 1962, General Greer's concern had put down roots more substantial than an academic distrust of land recovery as a technique. The Gambit system was then essentially 500 pounds over design weight, and most of the overweight derived from complications introduced by the land recovery requirement. Moreover, the reasons for distrusting air-sea recovery modes had become much less valid since 1960. Successful Corona recoveries were proving to be less difficult as time passed. The anticipated danger of losing a capsule to a Soviet trawler or submarine had largely dissipated; overwater recovery plans contained provisions for dealing with virtually every imaginable contingency, while the possibility that a slightly misdirected land-recovery capsule might descend in either Canada or Mexico—or might drive into some populated area of the western United States—had not diminished.

Equally important, the disabilities arising from land recovery had not been appreciably lessened in the intervening years. Indeed, in many respects they had come into sharper focus. Over-water recovery, as developed in the Corona program, seemed a very simple process when compared to the planned land recovery scheme. In its descent toward the sea, a Corona reentry vehicle could safely shed all sorts of accessories—hatch covers and the ablative cone being the most obvious. Such jetsam fell into the ocean without danger to anything below, and then sank into the secure obscurity of a cluttered sea bottom. A land recovery vehicle could shed nothing that might come to earth as a lethal projectile or which, if discovered, might breach the security of the satellite reconnaissance effort. Everything that re-entered with a land-recovery vehicle had to remain with it. Finally, experience with the E-5 and E-6 reentry vehicles, and particularly the latter, did encourage great optimism about the feasibility of recovering bulky, weighty capsules.

Toward the end of July 1962, General Greer again raised the question of the desirability of land recovery with Undersecretary Charyk. Although still dubious, Charyk agreed that Greer should look into alternatives. In point of fact, the most desirable alternative had occurred to Greer some days earlier. Thoroughly familiar with Corona, he had concluded that it might be entirely feasible to modify the Gambit vehicle to accept a Corona-type recovery capsule. After mulling over the idea, he decided that was the sensible and logical way out of the current dilemma. Having broached the thought to Charyk and gotten agreement that the idea had merit, he went directly from Washington to the General Electric facility in Philadelphia. Hilliard Paige, GE's senior satellite program official, was absent when Greer arrived, so the general settled down at Paige's desk and wrote a longhand memo authorizing GE to do a quick study of the feasibility of "gluing the Discoverer capsule on the front end of Gambit."[29]

Encouraged by preliminary findings, General Greer induced Charyk to formalize the inquiry. On 28 July the undersecretary agreed that a major policy decision was necessary on "the question of land recovery in the entire satellite reconnaissance program." He acknowledged doubts about the wisdom of relying too fully on " . . . the more complex and larger vehicles being developed toward land recovery" and asked for a "white paper" on the merits and shortcomings of land recovery. He also suggested a study of the feasibility of designing a capsule which could be retrieved in any of several different modes as individual mission circumstances dictated. Later that day Charyk added a requirement for a broader study of simplifying Gambit to provide "possible alternative modes of operation" including sea recovery.[30]

Apart from Greer himself, at least one other senior officer in the West Coast establishment had given serious thought to the land recovery problem during the early summer of 1962. Colonel Paul Heran, director of the E-6 program (which was then entering its flight test phase and had begun to encounter problems in recovery techniques), had looked into land recovery options for his satellite and had concluded that while the technique was feasible for E-6 it was not particularly attractive. Charyk was familiar with this conclusion as well as with Greer's severally expressed reservations. On 30 July 1962, General Greer discussed the Corona-capsule idea with Colonel J. L. Martin, Charyk's principal staff officer for the satellite program, in Washington. By the start of the following week he had received an advance report from GE, in Philadelphia. He promptly advised Martin and Charyk that a relatively minor modification to the Discoverer capsule would provide "a vastly simpler scheme for recovering record data on certain special projects." He asked for authority to start GE on a full investigation and, if there were no technical obstacles, to buy and modify sufficient capsules for the Gambit launches.[31]

In the meantime, Greer had assigned to Colonel Riepe the task of responding to Dr. Charyk's formal query of 28 July. Riepe was cognizant of Corona, but the Gambit people who worked for him were not. Moreover, they were, like all good project people, convinced that their current approach was best. Thinking chiefly in terms of modifying the current Gambit capsule for air-catch recovery, as had been suggested—and dismissed as "impractical"—six months earlier, they displayed neither optimism nor enthusiasm. They pointed out that the current deceleration parachute was totally unsuited to an air-catch operation and that the capsule had been designed to sink if it came down at sea. The heat shield was specifically designed for land impact, as was the basic structure of the recovery capsule itself. Moreover, the command and control system intended for Gambit was integrated with a capsule design built around the philosophy of small dispersion errors; it would be unsuitable for a capsule susceptible to relatively wide dispersion. Then there were considerations involving facilities: negotiations for the command and control station had been pressed (though with no particular success), and a start had been made toward the construction of buildings on the Wendover range, in Utah.

On the other hand, more than 600 pounds of orbital weight could be chopped by going to an overwater recovery mode. The elimination of the land recovery requirement would also permit earlier testing with less risk, would reduce requirements for orbit adjust, and would (at least in theory) enhance the probability of recovering film, since over-water recovery techniques were by then well proven. Such reasoning was based on the little that Gambit people knew about the details of the Discoverer recovery vehicle.[32]

Some among General Greer's people believed that land recovery should be continued at almost any cost, considering its eventual adoption inevitable. In his summary study, forwarded to Charyk on 4 August 1962, the general took the opposite view, noting that many of the original motives for developing a land recovery capsule had been invalidated by the passage of time. The enormous expense of maintaining a sea recovery force to back up air catch operations had been a point in favor of land recovery in 1960, when the first Discoverer recoveries were successful, but with the refinement of air catch techniques the need for such elaborate surface forces had disappeared; air catch with limited ship and frogman backup had been well proven in the previous several months. The danger of capsule capture, the probability of loss, and the logistics of an air-catch technique had become less significant and techniques improved, while development of a land-recovery capsule had underscored new problems: weight, complexity, reliability, and performance penalties. In retrospect, the disabilities of land recovery seemed to have overtaken any earlier advantages. So, said General Greer, it was his conviction that the mobile air-sea recovery mode was "far simpler and has overwhelming operational advantages over fixed base recovery."* He predicted that the continued evolution of guidance

---

* Such opinions were supported by extensive comparison tables which showed the air-sea techniques to be best on the basis of all possible considerations. Land recovery showed up as the least promising technique, with island recovery next lowest. Among the factors considered in the study were all-night and all-weather operations, the dangers of dropping a capsule into a neutral country, requirements for precision, the status of several recovery techniques, the use of multiple capsules, the ability to conceal a recovery operation, and hazards to population.

systems would further reduce logistic requirements and increase accuracy, making multiple recoveries feasible in the process. (The general was also thinking in terms of a capsule suitable for use in any of several recovery modes—land, island, sea, and air-catch—with fewer specialized requirements than the contemporary land recovery designs.)[33]

On 6 August, Captain Frank B. Gorman (USN), General Greer's plans chief, summarized the status of various recovery techniques in a special presentation to Undersecretary Charyk. Almost simultaneously, General Greer was mulling over a problem of funding which bore directly on the newly pregnant question of continuing land recovery plans; he recommended on 14 August that requests for facilities' funds for the Wendover range be withdrawn from the fiscal 1963 budget totals, concluding that they would be extremely difficult to defend in the existent climate. Ten days later, Colonel Riepe made a separate presentation to Charyk on recovery matters and at its conclusion received instructions to plan for initial systems tests over the Pacific rather than over the Wendover range. The undersecretary also decided that development activities related to the original land-recovery capsule should be reduced to a minimum expenditure rate, accepting the probability of program slippage if it were later reinstated.[34] Separately, Charyk authorized Greer to begin immediate development of a Corona-type recovery system for Gambit, planning on a June 1963 first flight date.[35]

For two weeks following Charyk's 24 August decision, both the new and the old approaches to recovery were kept in being. In that interval, the undersecretary consulted with CIA members of the National Reconnaissance Organization. The situation was complicated by the fact that all matters involving use of the Corona capsule were in CIA custody; the agency maintained a jealous control of Corona security, opposing as a matter of policy all proposals for broadening the dissemination of information on the Corona recovery system. Although to many observers CIA's caution seemed to verge on the psychotic, there was no denying that the use of the "Discoverer" capsule system in a non-CIA reconnaissance system would increase the chance of compromising Corona. Memories of the U-2 incident were too fresh to encourage any laxity. In 1962, the consequences of a disclosure that the "science oriented" Discoverer program had always been a CIA-sponsored reconnaissance scheme were too frightening to contemplate. The agency therefore insisted from the start that any provisions for using the Corona capsule configuration in Gambit had to begin with thorough protection of all aspects of Corona security. General Electric, which made the "bucket" for the CIA, agreed that use of the Corona equipment in connection with Gambit could easily jeopardize the four-year-old cover story unless some means could be concocted for concealing the origin of the capsule.

On 18 September, General Greer and Undersecretary Charyk met in Los Angeles to settle the main question. Charyk had by that time come to the viewpoint that land recovery was a sophistication of reconnaissance techniques which, though highly desirable, might take another decade to perfect. He still felt that operational costs, system efficiency, and security would benefit from land recovery, but he agreed that it was not immediately essential or feasible. General Greer commented mildly that land recovery was a useful emergency capability, but one not necessary in the current situation. He added, as an aside, that he had never firmly believed that the land recovery mode would be used for the first Gambit. The need to recover was too compelling to risk the additional complications of an entirely new technique when a proven recovery system was readily available.

Charyk capitulated, approving use of the H-30 (Corona) capsule on the first ten Gambit shots and withholding a decision on later launches. He authorized cancellation of current studies on precise land recovery but added the proviso that Greer should undertake a study of modifying the H-30 for land recovery as an option to be considered after the first ten firings of Gambit.* The approved plan, then, involved adapting the original H-30 to the first ten Gambits and developing a mildly modified H-30, with a capacity for emergency land recovery, for later use.[36]

General Electric suggested several possible ways to hide the fact that Corona equipment was being used in Gambit and on 18 September, the day of the actual decision to proceed along the Corona capsule route, CIA recommended adoption of one of these. To the Greer people on the West Coast, the GE-CIA recommendation seemed unduly complex; they proposed a compromise. On 28 September, CIA agreed. The final procedures (which were complex because they were designed to keep knowledge of Corona from the Gambit people who would be using

---

* Negotiations for a tracking station site had to be terminated once the need for land recovery had disappeared. An amicable break off of discussions was complicated, unhappily, by one of those periodic lapses in communication that troubled the government. The project office believed that conversations through the State Department, had been discontinued in August. It later developed that several agencies involved in the affair had not been advised that the need for a station no longer existed. As a result, a formal agreement was very nearly signed before the kinetic energy of the original conversations could be harmlessly drained away. Final orders cancelling the station proposal were issued on 20 September 1962. A further refinement of Gambit station requirements, in February 1963, led to the abandonment of earlier plans for using the Annette Island site (in Alaskan waters) as a Doppler radar tracking site.

the device) provided for design of the H-30 for Gambit within GE's Chestnut Street establishment and the performance of qualification tests at Valley Forge. All of the "white" components would be made and tested at Chestnut Street and all "black" work conducted at Valley Forge. The purpose of the arrangement was to keep non-Corona people from learning that for more than three years the CIA-purchased Discoverer series of satellites had actually been carrying reconnaissance devices.[37]

The effect of the transition from a land-recovery system to the Corona recovery system included a slippage of at least one month in launch schedules. At that cost, the very troublesome weight difficulty that had earlier afflicted Gambit was eliminated, the complexity of the early design was materially reduced, and the requirement for a separate recovery force within the continental United States could be cancelled. On the whole, it seemed a worthwhile exchange.[38]

Dr. Charyk was not entirely happy with the outcome, however. His reluctance to abandon land recovery as a Gambit objective almost certainly stemmed from his original commitments to that mode during the Samos realignment period of 1960 and from the fact that the President and the National Security Council had been encouraged to expect a land-recovery system to become operational during 1963. Although he accepted the inevitability of the change, he never displayed any special fondness for the thought that the original Gambit concept had been modified. Greer, more pragmatic, was well pleased with the course events were taking. Not merely had the uncertainties of Gambit film recovery been reduced by the change in recovery techniques, but it appeared that the whole of the Gambit system had been markedly simplified.[39] In the wake of E-5 and E-6 experience, simplicity was a virtue for which he had a marked respect.

Prompted in part by hard questioning during a meeting with the Foreign Intelligence Advisory Board and the "special group" of the National Security Council, Dr. Charyk in early October 1962 questioned the adequacy of project-office-level management. Charyk characterized Gambit as "imperative" and urged that it be pressed with a "maximum sense of urgency, noting that the "extreme political sensitivity of any other method of obtaining such photography" made it essential that "no reasonable steps should be omitted to guarantee its success at the earliest possible time." Discouraged about the rate of Gambit progress Charyk specifically suggested to Greer the appointment of an extremely able project manager and the start of an exhaustive technical review to locate any problems remaining in the Gambit program. Resolution better than the two-foot requirement of 1960 was desirable,

he emphasized. He also cautioned that money was not unlimited and that greater management talent rather than more funds should be applied to the program.[40]

In all probability, the prevalence of over-runs, particularly at General Electric, the threat of new schedule slippages, and the increasing cost of the Gambit program prompted Charyk's sudden outburst of concern for the validity and future of the development. Such factors certainly were at the heart of his indirect suggestion that Colonel Riepe be relieved—a suggestion that he separately discussed with General Greer by telephone. Charyk, who had made efficient management his fetish at the time he acquired custody of the satellite reconnaissance function, tended to ascribe most of Gambit's contemporary difficulties to deficiencies in management at the program office level. He was particularly concerned at the possibility of further schedule slippages since Gambit offered the most promising approach to the task of discovering, at any given time, whether the Soviets were actively preparing their military forces for use.

The coincidence of Charyk's anxiety with the start of the Cuban missile crisis of 1962 could scarcely be ignored; even though the United States did not have clear evidence that Soviet nuclear-warhead missiles were being emplaced in Cuba until the second week of October, concern for that possibility had been mounting since the previous August. Obviously, it would have been much easier to deal effectively with a Soviet missile threat in Cuba if the administration had detailed information on the degree of Soviet preparation for quick use of strategic striking forces. This, then, was a central consideration in Charyk's desire to accelerate Gambit progress and to improve the quality of Gambit products. The prospect of program delay rather than acceleration, and of photographic degradation rather than improvement, certainly influenced him to suggest the assignment of a new program office chief.[41]

General Greer, who had to decide the fate of both the program and its immediate manager, was scarcely indifferent to the circumstances that had moved Charyk to such a position. The E-6 program was in grave technical trouble in October 1962, having experienced four failures in as many flight attempts. Nor was there available any convincing evidence on which to base a rebuttal of Charyk's stand. Through a succession of misfortunes mostly originating in the pre-1961 Samos program, it had been necessary to cancel each of the major photo-oriented reconnaissance programs originally assigned to SAFSP except E-6 and Gambit. And E-6 had taken on a distinctly unhealthy cast. True, the most obvious defect in Gambit design had been eliminated with the decision to adopt air-catch recovery techniques and the Corona recovery vehicle.

But the prospect of program slippage because of faulty attitude control development could not be banished and there was no ready means of insuring that the rather complex Gambit camera system would function with complete propriety during its early flight trials.

On 5 October, Greer told Charyk "with some reluctance" that the most certain way to strengthen Gambit management along the lines Charyk had indicated would be to transfer official custody of the program from SSD to SAFSP.* He reaffirmed his desire to keep Colonel Riepe in charge of Gambit development. Moving the program into SAFSP, he told Charyk, would give the development the benefit of prestige that adhered to any effort identified with the secretary's office, although it seemed possible that identification of Gambit with reconnaissance objectives might follow. In Greer's eyes, that was not a disqualifying handicap. He reminded the undersecretary that the United States had constantly maintained the basic legality of satellite reconnaissance under international law and that the nation had never denied either the existence or the employment of orbiting camera systems. The chief purpose of concealment now, he suggested, was to cloak the scope and operational success of such operations. That much could be done within SAFSP. The remote possibility that national policy might shift, in which case it would be difficult to continue any effort even indirectly associated with reconnaissance objectives, was the chief argument against moving Gambit.

The general was not optimistic about the prospect of improving the quality of Gambit photography, at least in the first several flights. He told Charyk that the original definition requirement, two to three feet in resolution, would very probably be satisfied—though he observed that not all experts agreed with him on that score. The mirror was the critical item, being chiefly responsible for both distortion and light loss which reduced resolution. Greer cautioned that results from the first few flights might not bear out his conviction that Gambit would prove itself; past experience with "new" space vehicles (into which category the General Electric orbital control vehicle certainly fell) was not such as to encourage undiluted optimism.

As for priorities and emphasis, General Greer noted separately that "...it is difficult to convince either contractors or military personnel involved in administration of this program that it enjoys any special priority or importance. The one infallible indicator of status—timely and adequate funding—is and had been consistently absent since the turn of the fiscal year."[42]

Although General Greer had essentially reacted to Charyk's message of 3 October by defending the status quo, he thereafter set afoot major changes which, within 60 days, markedly altered both the configuration and the character of Gambit. On 30 October he announced to members of the 206 program office that Colonel William G. King was assuming management responsibility for their project and that Colonel Riepe had been detailed to a new and demanding SSD program.[43] Four weeks later, Colonel Riepe was officially named director of Program 437, identified by Air Force headquarters as an extremely high priority project aimed at the early development of a useful satellite interceptor.[44] General Greer had earlier discussed SSD's need for an experienced space program manager with SSD's commander, Major General B. I. Funk, and with Colonel Riepe's knowledge had worked out a transfer arrangement.[45]

King, who had been intermittently associated with satellite reconnaissance for nearly 10 years, had special qualifications for the Gambit assignment. As Samos program director during 1959 and 1960, he had been a participant in the bloody infighting that accompanied the Samos reorganizations. He had been one of the first to recognize the superiority of film recovery over readout techniques for Samos and had been notably outspoken in his support of recovery as a technique. Since cancellation of the E-5 program the previous winter, he had served mostly as a special plans officer for General Greer, conducting detailed studies and comparative analyses of the various systems proposed and in development, although he had also retained responsibility for the slow-paced Valley program, an early effort to develop a search system with resolution potential.

Admittedly, the Gambit program was a bit out of hand when King took it over. Riepe had reacted to Charyk's continued pressure for insuring a first flight success by creating an elaborate test regime which had substantially increased the cost and complexity of the development. There were indications that Charyk did not have a high regard for Riepe's ability, General Greer, who thought well of him, nevertheless conceded that in his dedication to the assignment Riepe had tended to overelaborate the program and the program office. In the circumstances of October 1962, simplification of both seemed necessary. King's job, then, was to devise and put into effect measures for restoring full confidence in program success—a commodity not always abundant that fall.

Immediately after taking over the program, Colonel King discovered that the adaptation of the Corona capsule to Gambit uses had gone thoroughly off course. The situation had its origin in a series of

---

* The desirability of shifting 206 into the SAFSP structure had been examined in some detail as early as July 1962. By October, Colonel J. W. Ruebel, General Greer's special assistant, had worked out the basic details of the transfer plan later adopted and had composed a "rationale" for public consumption.

basic misunderstandings complicated by a lack of knowledge in the program office and among GE and EK engineers.

Greer's original intent, confirmed by Charyk, was to "glue on" the Corona recovery vehicle. Elaborate or extensive modification of either the capsule or the orbital midstructure was neither intended nor desired. Because of the rigid compartmentation of programs, however, only Colonel Riepe among the Gambit program office people had a reasonably full knowledge of the Corona program. Corona provided two years of carefully concealed experience with unpressurized operation. Reasonably enough, lacking any indication that unpressurized operation was possible, Gambit people concluded that pressurization of the film cassette, a basic feature of Gambit, would have to be continued in the new recovery capsule. The chief difference between the two cassettes, once pressurization requirements had been sorted out, lay in the greater film width of Gambit, a factor that General Electric's engineers must certainly have taken into account in responding to Greer's July inquiry on the feasibility of converting Gambit to a Corona recovery vehicle.

In the course of changing over from land recovery to air catch, the Gambit office had eventually authorized General Electric to develop a recovery vehicle essentially capable of accepting the original—pressurized—Gambit take-up cassette and film chute. Because the unit was substantially larger than the Corona cassette film chute package, General Electric had scaled up the Corona capsule, making it deeper and increasing its base diameter. Such changes presented the program office with what was neither a Corona nor a Gambit capsule, but something resembling the former in external configuration while being nearer the latter in overall size and internal arrangements. It was, for practical purposes, so significantly different from the original Discoverer capsule as to require proof testing. Colonel Riepe's inquiry about the procedures to follow in scheduling a drop test program first brought the matter of capsule configuration forcefully to the attention of General Greer's immediate staff. Obviously there was no justification for drop-testing Corona capsules. There was a good case for the argument that no need for such a major modification of the Corona recovery capsule could be demonstrated, that pressurization—which was the cited justification for the modification—was entirely unnecessary.

As it happened, one of Kodak's senior people in the Gambit program was aware of the fact that the requirement for pressurizing any part of an orbital camera system had long since been invalidated. F. C. E. Oder, deeply involved in the early Corona effort, had retired from the Air Force and joined EK. Imbued with the security consciousness of the Corona activity and no longer active in that program, he did not consider himself entitled to pass his information to fellow workers at Rochester. Major John Pietz (of Greer's staff) solved that difficulty by flying to the New York plant and briefing a select few EK people on Corona. Thereafter, EK could work on an unpressurized cassette design with some confidence that it would not fail because of the lack of a pressurization feature. But by this time (mid-October 1962), GE was well along in the final design of a bigger Corona-style reentry vehicle. A full-scale mock-up had been built, substantial sums added to the contract totals, and an extensive test program planned.

One of Colonel King's first moves after moving into Gambit management was to advise General Greer that he thought the design of the adapted capsule represented much more of a change than Greer had intended. Greer, who had ordered that changes to the Corona capsule should be minimal, was disconcerted. He forcefully endorsed Colonel King's suggestion that the original intent of the modification be reinstated and that the rapidly burgeoning General Electric development effort be stopped in its tracks. King met with key GE officials two days later and defined the objective of the capsule change in terms of General Greer's appreciation of the need. Cross-briefing of Gambit people on Corona, a continuation of the process earlier begun at Eastman Kodak, eliminated any excuse based on technical uncertainty.

The episode apparently was partly the consequence of a semantic gap. Colonel Riepe and the 206 program office people considered that the modifications General Electric proposed to make were "minor" in the terms of General Greer's original instructions. The problem was compounded by a general lack of Corona information among Gambit people, both in the program office and on contractor staffs. If pressurization was either essential or highly desirable, a "big" capsule was inevitable, and in the absence of knowledge to the contrary it was not illogical of Gambit program people to continue to believe in a need for pressurization. In Colonel King's view, the new GE version of the Corona capsule represented a sharp departure, a more-than-minor change. General Greer, it developed, agreed with King.[46]

The main issue was finally disposed of early in December 1962. Stating his preferences plainly, General Greer told Colonel King: "The name of the game is to adopt it [the Corona capsule] for 206 without introducing a change external or internal which will result in failure on the first try or otherwise prejudice its reliability." King responded with the advice that he had imposed an "absolute minimum" change policy and that earlier changes in the external configuration had arisen from too strict interpretation of instructions that adoption of the Corona recovery system was to have a minimum effect on the payload.[47] As it happened, the payload was the least risky element of the system, command and control representing the most difficult.

By that time, Colonel King had made it entirely clear to both GE and EK that system changes were to be minimal. Deviations from the original external configuration of the Corona capsule had to be cleared personally by King before approval. By all indications, the external changes would be slight. The general policy, King added, was to use flight-proven components wherever possible, keeping all change at a minimum but altering the details of payload configuration as essential to the requirement for limiting external change.[48] Early in November, while Colonel King was in the early stages of sorting out the technical complications of Gambit and subjecting them to a detailed analysis, General Greer reactivated the suggestion of transferring the entire 206 program to SAFSP. Answering earlier objections, he explained to Charyk that such a move did not imply "surfacing" the development and acknowledging its reconnaissance objectives, that the payload would remain covert and procurement "black." Moreover, the cover plan devised in SAFSP promised to perpetuate the legend that 206 ("Cue Ball") was in some way related to a bombs-in-orbit program. The explanation for transfer from SSD to SAFSP need not be either complex or particularly sophisticated; a straightforward statement that because of program priority it was being put under the direct control of the Secretary of the Air Force would satisfy those unknowing that covert programs were conducted within the Air Force. Greer expected that others aware of the existence of clandestine activities would deem it unthinkable to move a concealed reconnaissance program into a reconnaissance organization and would be more firmly convinced than ever that 206 had some mission other than satellite reconnaissance.

"Children or half-wits, if they care, will most likely reason directly to the correct deduction, i.e., if it's assigned to SAFSP, it's reconnaissance. In as much as we will do nothing to confirm this," commented Greer, "and we will insure that some actions are apparently inconsistent with this hypothesis, I think there is a good chance of fooling-or at least confusing—the professional espionage agent, who is presumably neither a child nor a half-wit."

There was another consideration, which General Greer did not specifically identify in his correspondence with Undersecretary Charyk but which certainly influenced his judgment on transferring Gambit to SAFSP. A Department of Defense Directive—DOD Dir 5200.13—originally published in March 1962 and revised later that year, had placed all military space programs in a "no publicity on payloads" and "special access, must-know" category. Individual access lists were to be maintained for each program and information on each program was to be confined to those having been granted a "specific need to know" recognition. Random numbers were substituted for the previously used popular names and launch announcements were restricted to a bare statement of the type of booster and the date of the operation. In those circumstances, it was no longer possible to identify a "Samos" payload solely from the fact of launch security; all military space launches were to be conducted under tight security provisions. Thus it was increasingly difficult to come by information about most Air Force space programs (though in point of fact, 5200.13 was rather casually administered within the fraternity of those having general access to classified information), and to a degree all cover stories had become somewhat redundant.[49]

The arguments were effective. By 20 November Charyk had concurred in the "desirability" of transferring 206 to SAFSP. Major General O. J. Ritland, who was then part of the Air Force Systems Command headquarters staff,* was called in to brief General Funk on the realities of the situation.[50]

That much out of the way, Greer and King set about changing the technical character of Gambit. Since the questionable stability of General Electric's orbital control vehicle currently was the most dubious aspect of the development, they conceived the idea of leaving the orbital vehicle attached to the Agena second stage through the whole of the first mission. The Agena, which had a generally reliable stabilization and control system but one considered insufficiently precise for Gambit operations, could stabilize the Gambit camera long enough to secure operating experience and proof of system feasibility. Greer and King, with memories of E-5 and E-6 cancellations caused by on-orbit failures, were determined that the first Gambit flight should return at least one good picture. That achievement in the E-2, E-5, and E-6 projects might well have insured their continuance, at least temporarily. Greer was adamant that nothing

---

* In 1962, the Air Force Systems Command embarked on a determined campaign to return control of satellite reconnaissance to "normal" Air Force channels. "Loss" of Gambit represented a defeat in that campaign. Charyk later quashed the whole activity, but it experienced a brief revival after his departure in early 1963.

of the sort would be said of Gambit. Charyk was not convinced that the "one picture above all" outlook was the correct one, but it seemed possible that he could be brought around.

There was more to "hitchup," as the notion of keeping the orbital control vehicle attached to the Agena was called, than met the unwitting eye. An elaboration of the scheme involved use of the roll-joint coupling invented for Lanyard. Should the orbital control vehicle prove generally unreliable, it might be possible to introduce the Lanyard roll joint between the Agena and the payload end of Gambit, eliminating reliance on the stability and control elements of General Electric's orbital control vehicle.

On 29 November, General Greer took the results of a preliminary analysis of the "hitchup" and "roll joint" ideas to a meeting with Undersecretary Charyk. The undersecretary showed interest. On his return to Los Angeles, Greer drafted an authorization for continued study of these options and sent it to Washington for endorsement. Later that day (30 November) the second major change to Gambit in two months was tentatively approved.

The chief difficulty in the latest idea was devising a non-compromising means of bringing the roll joint part of the technique into the Gambit program. As was the case with the Corona reentry capsule, the roll joint was quite unknown to most Gambit people and because of the security compartmentalization that existed within the reconnaissance program structure it seemed highly unwise to disclose the existence of Lanyard to large numbers of Gambit workers. So "Charyk's" message of 30 November, actually written by General Greer, contained the "suggestion" that Greer contact Lockheed about the roll joint as "... he [Charyk] believes a similar idea was once proposed and possibly designed in connection with another space program."

The kernel of the cover story here outlined was that Lockheed would be empowered to "develop" the earlier "idea," delivering finished roll joints to Gambit as though they were new items with no relationship to any other reconnaissance program. The scheme was so simple it seemed foolproof.[51]

In his conversation with Dr. Charyk on 29 November, General Greer had promised that additional measures for insuring the success of the first Gambit flight would be proposed in the course of a fullscale program review on 14 December. On that date, he, Colonel King, several program people, and a team from Aerospace Corporation not only reviewed the program but proposed still another technical innovation. (Charyk had earlier approved a contract technique change which eased the financial pressure on the first six flight vehicles, agreeing that they could be purchased on a cost-plus-fixed-fee basis with the seventh and later Gambits being funded on an incentive fee basis.) The latest change provided for incorporating "Lifeboat" provisions in Gambit. "Lifeboat" was another technique originated in the Corona program; it involved the provision of independent reentry command circuitry (including a receiver), a separate magnetometer, and its own stabilization-gas supply. All were independent of the main systems. If the primary reentry systems became inoperative for any reason, "Lifeboat" could be separately actuated. The magnetometer used lines of magnetic force around the earth as a longitudinal stabilization reference, permitting the device to place the Agena (or any other suitably equipped orbital vehicle) in a proper attitude for the start of de-boost, relying entirely on its own gas supply for attitude control and a taped command sequence for the recovery process. In several experiences with Corona vehicles, "Lifeboat" had proved highly reliable.

On 19 December, the undersecretary formally authorized the "Lifeboat," "hitchup," and "roll joint" expedients for Gambit. "Lifeboat" was to be a permanent part of the total system, "hitchup" was to be incorporated in the first four vehicles (but a determination on use would be made on a flight-by-flight basis), while "roll joint" was to be developed "as a bona fide operational substitute for the OCV [orbital control vehicle] roll system."[52]

At the time that these additions were made, General Greer approved a proposal by Colonel King to delete rather substantial portions of the elaborate test program earlier scheduled. There was no real alternative if the launch schedule, now specifying first flight in July, was to retain any validity. Both King and Greer were uncomfortably aware that reducing the number and scope of development tests was risky. They were also aware, however, that another contract overrun or a new schedule slippage represented an equally grave danger to long-term program stability. Experience—not always the best criterion, but in this instance the only one available—seemed to indicate that the simplest and most direct technical approaches worked best for reconnaissance satellites. Although it was possible that pruning away tests might lead to the failure of some major component, the program innovations of September-December 1962 provided considerable insurance against a major catastrophe. Air-catch and use of the H-30 capsule overcame objections to the faults of the original Gambit recovery technique, and Lifeboat provided greater assurance of recovery success. Hitchup represented a feasible means of increasing the probability that the camera system would have a chance to display its abilities without succumbing to the frailties of an unproven stability and control system. Roll joint was a safeguard against the long-term unsuitability of the GE orbital control vehicle. On these counts, Gambit was a much more realistic program in late December 1962 than had been the case four months earlier.[53]

Finally conceding that the most vital initial objective of Gambit was to return one good picture (Greer's frequently stated goal), Dr. Charyk nevertheless insisted that all flights subsequent to the first had to be programmed to return useful pictures of pre-selected intelligence targets. He specifically rejected the concept of a step-by-step approach to an operational configuration through research and development improvements. His philosophy was the key to the reason for incorporating the roll joint development:* if it were necessary to rely on roll joint because of the failure of the GE orbital control vehicle, the GE effort could be discontinued. Degradation of picture quality was a probable consequence, but the degree of degradation could not be accurately estimated in advance. Theory had it that the greater delicacy and precision of roll obtainable through the GE vehicle was essential to Gambit operation. The roll joint system could provide only 100 stereo pairs of pictures of selected targets during a single mission, about one-third to one-fourth of the current expectation for the GE vehicle, and one-sixth of the original requirement. Roll joint was designed to permit shooting at angles as great as 30-degrees from the vertical, with intermediate settings every five degrees. In late 1962, Gambit experts were not optimistic about the prospect of compensating for smear and image-motion-compensation errors when roll joint was in use; in the event, more than a year later, experience was to show that they had been far too pessimistic about the stability capabilities of the Agena and too exacting in their requirements for camera stability in Gambit.[54]

---

* Roll joint development, from a technical standpoint, presented no special difficulty. It existed, and worked, as part of Lanyard. But getting it from the Lockheed Corona works to the Gambit assembly building was something of a problem. Contract authority originated in a change order to the existing Lanyard contract. The basic pattern devised to get the H-30 capsule from Corona to Gambit channels was adapted thereafter. Fabrication was "in the white;" assembly, test, and qualification were "black" processes. Aerospace people were told only that Lockheed had proposed the roll joint to Greer somewhat earlier and that the firm was so well advanced in a hardware sense, because of such earlier work. Thus there would be no need to brief any new Aerospace people on Corona. Some Lockheed people would need Gambit briefings, but that was inescapable. Documents and correspondence relating to roll-joint origins would remain in Corona channels until delivered to SAFSP, at which point all references to the Lockheed organization and to Corona associations would be deleted from the documents. Thus "sanitized," the paper could go into Gambit channels. Risk of compromising Corona was negligible.

The first 19 Gambits would cost money, of which most had been expended by December 1962. (Some more was budgeted against flight and operation costs through June 1965.)[55] In such a context, measures which would hold down costs were in order. There was no ignoring the need for program insurance applicable to several aspects of the total program. General Greer's instructions to King in October had emphasized three critical goals: staying within the budget, staying on the schedule, and obtaining one good picture. The prospect that Lanyard could fill the role earlier reserved for Gambit was but one reason for discernible concern at the possibility that Gambit might be terminated if it encountered major difficulties. In the reconnaissance program, none of the ordinary reasons for avoiding program cancellation situations had much weight. No embarrassment could result because news of a cancellation would be confined to a small circle of familiars. Neither the public nor Congress would have occasion to carp at a program cancelled, a factor of some importance to the normal program management structure and one of the more important influences in continuing programs which in a totally rational world would probably have been terminated. Either budget overruns or significant schedule slippages could degrade Gambit's prospects.

By early January 1963, Colonel King had pared off money in fiscal 1963 and in fiscal 1964 program fund requirements. (The total largely represented cuts in engineering work not essential to program success, the purchase of fewer spares, a reduction in the requirement for post-flight data reduction, tightened quality control, elimination of much documentation earlier called for, simplification of reentry vehicle tests, and a reduction in qualification tests.) King proposed that program funding be further reduced by eliminating some of the work earlier scheduled, and suggested to General Greer several areas which seemed ripe for attention. Some additional funding might be gained by cancelling parallel efforts and alternative subsystem developments.

Although Greer was attracted by the possibility of reducing program costs, he was reluctant to adopt all of the measures suggested by his new Gambit manager. Although agreeing that it was feasible to cut a rather extensive vendor reliability program "to the minimum essential" and to employ an earlier scheduled ground test system in the flight program, he was extremely dubious about the wisdom of halting work on an alternate horizon sensor since there still was some uncertainty about the utility of those then programmed.* He expressed similar reservations about King's proposal to halt work on an alternate velocity meter. But in both instances, he told the colonel, if it appeared that funding the work into an indefinite future would cause invalidation of established budget ceilings, he would be quite willing to reconsider his decisions.[57]

Unexpected funds problems rose to plague Gambit shortly after the question of what to cut and how deeply had apparently been resolved. In reviewing his budget, Colonel King discovered that the Agena cost estimates given to Dr. Charyk in December "were grossly low." The total of costs for the modified program would be higher than planned, chiefly because of configuration differences which had not been adequately weighed in costing the adoption of Agena vehicles based on the Discoverer program. Because of the earlier budget overhaul, however, the deficiency was substantially less of a problem than it might otherwise have been.[58]

Funding difficulties were scarcely unique in satellite reconnaissance. There were times, however, when Gambit seemed to have its own special affinity for such difficulty. Colonel King remarked at one point, early in 1963, that the program had been stigmatized as a high cost development and GE as a high cost producer—and that both characterizations might be justified. He told General Greer, "It is my feeling that we may have outlined a program to the contractor that is inherently expensive; our scheme of managing, reviewing, and presumably safeguarding a high degree of success is a big order to be swallowed. Certainly it isn't the cheapest way of doing all things. We may have built such a super foundation that we cannot afford the remainder of the structure."[59]

Yet despite such obstacles, which were both very real and very important, Gambit continued pretty much on schedule. In late 1962 the main uncertainties which affected expectations of operational success concerned the vehicle stability and film recovery aspects of the program. Those areas received greatest attention, through conversion to the Corona capsule, adoption of hitchup, roll joint, and lifeboat options, and the concentration of effort on horizon scanner development. In other respects, the program was doing quite well. The camera, in particular, seemed to be coming along nicely. Even before the stability expedients were adopted, camera operating tests had demonstrated an equivalent 2.7-foot ground resolution at better than 115 lines per millimeter. The only problem that appeared to offer any particular difficulty was the motor speed drive, and it was far from insurmountable. A mirror mounting problem, that had earlier given trouble (and which was similar to a problem then holding up Lanyard), had been essentially solved by November 1962.[60]

One additional change in the basic configuration of Gambit was recommended in January 1963 and approved for adoption on 28 February. This was a stellar-index camera, earlier treated as "purely an auxiliary package" but now considered quite important. The National Photographic Interpretation Center made the original recommendation, CIA's Herbert Scoville endorsed it, and Charyk approved its inclusion. The camera itself was to be that developed for the Corona-Mural. Because of procurement and installation delays arising from the advanced stage of completion of the first lot of Gambit payloads, the fourth Gambit was the first which could be scheduled to incorporate a stellar-index system. Each installation would cost.[61]

By virtue of circumstances, the fourth Gambit vehicle became the first in what was essentially a re-modified configuration. Hitchup capability was provided in all of the first six, but Lifeboat was an Agena installation in the first three, being shifted to the GE vehicle thereafter, and roll-joint capability was scheduled to be incorporated starting with the fourth system—as was the stellar-index camera. Early in March, funds for a total of 10 Gambits were provided (only six were then financed) and that month the earlier requirement for 19 was reconfirmed (but not completely financed) and an additional six Gambits were authorized to provide a quick-reaction and standby capability.[62]

---

* The horizon sensor problem, then more than a year old, was being attacked from three directions. The EK infrared system which had earlier been funded because it seemed to offer an attractive alternative to GE's dual-scan sensor had not proceeded as well as hoped. By December 1962 it was clearly a high risk development which promised rather less improvement over the GE system than earlier hoped. It was also, at that time, at a stage where major redesign seemed necessary if it was to satisfy original goals. Nevertheless, it still seemed to promise greater inherent accuracy than anything else available—if it could be perfected. The GE scan system was making satisfactory progress but in King's opinion would encounter both schedule slippages and overruns before 1963 ended. The third alternative, the Barnes system, promised to be lowest in cost and at least as good in performance as either of the others. King had urged Greer to approve cancellation of the EK program, continuation of the Barnes development and use of the Barnes scanner on the fourth and subsequent Gambit flights, and proposed to delay a decision on the GE program until the outcome could be more precisely estimated. General Greer agreed with the suggestion that the Barnes scanner be programmed for the fourth and subsequent Gambits but did not favor cancelling the EK effort for at least another month (after February) and felt that the GE system should continue to receive support. The uncertainty was not finally eliminated until September 1964, when Colonel King again urged adoption of the Barnes sensor. By that date both the GE and EK versions had proved their utility, but each was more costly and neither was better than the Barnes sensor. The Barnes model had other advantages; it was lighter, simpler, had the lowest power requirements, and operated over the widest temperature range of the three. After discussing the situation with Colonel King in some detail, General Greer agreed to such an assessment. On 17 September 1964, then, King notified GE that Barnes sensors would be furnished by the government for installation in the Gambit vehicle. A key factor in the decision was the ability of the Barnes sensor to function effectively in a winter environment, a feature which neither the EK nor GE models could satisfactorily demonstrate.[56]

The standby requirement had appeared rather suddenly, in mid-March 1963, although its origin could readily be traced to the Cuban Crisis of the previous October. The Gambit schedule in effect since early program approval had envisaged one launch every 40 days. By March 1963, the several high level agencies that maneuvered national policy in accordance with intelligence inputs had concluded that provisions should be made for launching a second Gambit in the event any primary mission was not successful. Second, the intelligence community saw a need for an emergency crisis reaction capability for the rapid launch of reconnaissance satellites. Once the basic need had been defined, what remained was a series of unanswered questions concerning potential: how much notice was required, what additional facilities would have to be built, would additional vehicles be needed, were there options on launch systems?

Colonel King's office responded to the original query of 12 March 1963 with the information that a backup capability could be provided by May 1964, that at least three and preferably four additional launch pads at Vandenberg would have to be modified to accept the Gambit-configured Agena, and that additional orbital vehicles were the critical items from a time standpoint. The answer was sufficiently comprehensive to permit the issuance of the 27 March order for six additional Gambits. (Authorizations for pad modification, which were included with the procurement approval, involved changing at least two launch pads to a dual configuration capable of accepting either thrust-augmented Thor-Agenas or Atlas-Agenas in a Gambit configuration.) The end objective of this program expansion was to satisfy plans for "a minimum intelligence cycle capability for coverage of the very highest priority special tasks, including only sufficient orbital duration to achieve the specific coverage..." Resolution at the level of the Gambit system, and in stereo, was desired. The problem, General Greer learned, "...is not to locate targets, but to inspect in detail activities at selected known targets."[63]

Further complications of the already complex problem thus outlined lay in a subsequent directive to study the use of tandem recovery capsules in Gambit to permit early recovery of a portion of the film while continuing the remainder in orbit until it was specifically needed. Corona was on the way to that configuration in 1963.

By early May, study of the problems of supplementary launch, standby, and quick reaction had been sufficient to show that a high launch rate could be maintained by keeping at least three pads in a Gambit configuration and by building up a modest stockpile of boosters and Gambit systems. The chief difficulties lay in procurement, since the only way to accumulate spares was to accelerate production rates or to reduce the frequency of launches. Either promised to be costly. There was some indication that fundamental changes in the Gambit program might ultimately be needed, including provision for both multiple recoveries and for an Atlantic Ocean recovery force.

Although the basic actions needed to provide a standby capability had been approved early in March, it was not until May that a persistent uncertainty over funding arrangements was eliminated. In some degree, the difficulty arose because Pentagon officials did not fully appreciate the intricacies of contract negotiations, particularly when they involved a manufacturer (General Electric) who apparently felt that the need for continued production was sufficiently great to elicit particularly generous terms. In order to keep the contractor responsive, General Greer had broken Gambit into blocks; General Electric was not aware of the total programmed and Greer did not propose to give the firm such information until the unit price became reasonable. Nevertheless, in order to maintain the required rate of production, Greer had to have in his funding reserve sufficient money to finance the complete approved production order. To accountants who looked at the General Electric situation without knowing these facts, it appeared that Greer's organization had a comfortable reserve of uncommitted money. In actuality, what seemed to be a "reserve" was required for technical contingencies and for satisfying vehicle requirements not yet formally on contract. Once these facts were made sufficiently plain, the funding difficulty began to diminish.[64]

Another pending uncertainty, involving the possibility of tandem recovery vehicles for Gambit, remained unsettled. The notion seemed feasible if certain rather substantial changes were made in the Gambit system and if a degree of photograph degradation were acceptable in the case of systems "stored in orbit" for considerable periods. But the cost was a bit high, reaching for development and for each vehicle. In general configuration, the proposed Gambit tandem system rather closely resembled a similar design evolved for SPAS-63 ("Spartan"), a Lanyard-category variant of the recently cancelled E-6.* None of its elements were technically unrealistic, though reliability might be low. In any event, General Greer's judgment was that nothing serious should be attempted in the matter of tandem-configuration Gambits until the original system had been well proven.[65]

By the time such matters had been resolved, attention was turning toward the impending first launch of Gambit. Booster-payload assembly had begun in February, after some delay because of the late arrival of prime components and the need to incorporate hitchup provisions. In order to protect schedules, Colonel King had agreed that it would be permissible to put the missing components into the total system during functional testing.

---
* See Perry History Volume I, Chapter IX

The command decoder had to be modified late in January 1963 to eliminate a conflict between command reception and execution. Subsequently, system tests disclosed the existence of some electromagnetic interference problems not previously suspected, but fixes proved possible without delaying the test program appreciably. The Agena for flight number one passed final acceptance tests at Sunnyvale on 21 March, slightly later than desired but still within the boundaries of the desired launch timing schedule. Early in April, Eastman Kodak located and corrected three sources of focus error in the camera system and later that month static tests of the assembled system showed resolution superior to that required of the Gambit system. (The focus problem was not critical for the first flight since it was used only during a roll maneuver, and hitchup was to be employed throughout the period of camera operation on the maiden flight.) All seemed to be going rather well.[66]

Then, during the late afternoon of 11 May, a faulty valve in combination with a deficient fuel loading sequence caused a loss of internal pressure in Atlas 190D, that being used in checking out procedures for the first Gambit flight. The booster collapsed on its stand, dumping both the GE orbital vehicle and the Agena on the concrete hardstand. The GE vehicle was severely damaged, the Agena to a lesser degree. Surprisingly, there was neither explosion nor fire, although 13,000 gallons of liquid oxygen and a full load of fuel sloshed over the stand and the nearby terrain. Equally fortunate, the payload did not split open, so there was no compromise of Gambit security. But the camera system was rendered permanently useless, a large part of the optics being demolished, and the recovery vehicle was so battered that further use seemed imprudent. Neither the camera nor the orbital vehicle were scheduled for the first Gambit test; the Agena, however, was supposed to be used in that launch.

Through a quick scavenging operation the program office secured an Agena to replace that damaged in the accident, using considerable overtime work to adapt the space booster to hitchup.* But hope for meeting the 27 June launch date lessened as the degree of launch stand and vehicle damage was assessed.[67] There was a possibility of further delay in a requirement from the undersecretary's office that the Agena and Atlas for the first Gambit shot be subjected to the same sort of "tiger team" pre-flight check that had resulted in a perfect booster operation record during E-6 launches the previous year. Both through SSD channels and in his own right, General Greer set up special review procedures to insure the basic reliability of the booster elements, assuring McMillan in mid-June that every conceivable means of assuring reliable operation had been employed. No program delay resulted.[68]

While boosters were attracting attention, final checks of the orbital control vehicle on a vibration stand uncovered subsystem faults which incapacitated the command programmer. Fixes and a re-test required two weeks, setting the launch date back to 10 July. As it happened, some delay was by then inevitable, the Agena modification to a hitchup mode having taken several days longer than predicted and pad repairs having continued until 22 May—which left too little time for a complete pre-flight checkout before the original 27 June launch deadline.[69]

One other set of developments had been continuing parallel to the technical aspects of launch preparation. These involved security and deception. For several months the process of closing off all avenues to information on the true mission of Gambit had been accelerating, not because the dense blanket of security and misdirection surrounding the project had ever been thinned in any degree, but because the extension of project activity to the launch complex at Vandenberg and the prospect of a recovery operation required a number of out-in-the-open arrangements and some measure of physical exposure. The CIA, for instance, was expressing uncommon concern about the possibility that the Gambit employment of Corona originated devices had "caused greater deterioration to Corona security than was anticipated." CIA had always been extremely sensitive to any threat of Corona exposure, however slight it seemed; the Gambit episode appeared to attract even more attention on that score than usual. It was the CIA's position that all possible alternatives should be attempted before resorting to the practice of briefing more people on Corona.[70] To a lesser extent, Greer's organization operated on the same premise, acting in January 1963—for instance—to change payload analysis procedures so that no test controllers at Sunnyvale would have to be made cognizant of Gambit's actual functions.[71] But a great deal of observation and a fair amount of quiet inquiry had confirmed General Greer's assumption that most minor participants in launch and orbit control operations merely performed their assigned duties without even wondering why. The flight controllers, for instance, were largely content to monitor their

---

* Agena damage proved to be less than originally estimated. Undersecretary Brockway McMillan, who had replaced Charyk in that post in March 1963, initially directed that the repaired vehicle not be used in the early Gambit program but General Greer subsequently urged its reinstatement. He reported that exhaustive tests had disclosed almost no harm to the vehicle and that having been subjected to even more thorough checks than most Agenas it was qualified for its intended use. McMillan accepted the recommendation. The Agena in question was numbered 4701.

meters and take their required readings without asking whether the incoming telemetry originated in a camera subsystem or a beta-ray sensor.

Special provisions had been made at Vandenberg to cordon off the Gambit areas in the missile assembly building from other projects that shared the facility. Then the E-6 people, for instance, were told that "your program is extremely sensitive," and that they were not to let the 206 people get a glimpse of their payload. The gambit worked marvelously, convincing the E-6 workers that 206 was much less sensitive than their own effort.

Other people at Vandenberg who had enough experience with either overt or covert reconnaissance programs to be suspicious of Gambit apparently concluded that 206 couldn't involve reconnaissance because other efforts, such as Corona and Lanyard, were fully satisfying all needs. There was also a general impression among the photographic fraternity aware of Corona and the E-5, E-6 programs that it was not technically feasible to stuff a big camera-film package into a vehicle dependent on a Discoverer-size recovery capsule.[72]

There were significant indications that the original misdirection of 206 program documents had been spectacularly effective. High-level officers who were briefed on Gambit during the early months of 1963 generally confessed complete ignorance about the character of the payload (concurrently disclosing a most interesting lack of curiosity) or admitted to a belief that it involved precise de-orbiting of some sort of nuclear weapon.[73] There was no pertinent press speculation whatever.

One of the problems peculiar to pretending that Gambit was a non-camera project was that a certain number of Eastman Kodak people had to be at the launch stand during final checkout. The problem decreased appreciably when Lieutenant Colonel John Pietz and Colonel J. W. Ruebel ran a careful study of needs and discovered that no more than four or five camera specialists were actually needed. In dress rehearsals for the first launch, they were literally smuggled into the launch area in the back of an unmarked van. The practice was dropped, however, when the driver wrecked the empty truck while returning from one delivery run. Thereafter the needed specialists entered the launch zone as inconspicuously as possible, but using more conventional means of transport.[74]

There was some concern for the mode of getting recovered film from the mid-Pacific drop zone to the Eastman processing facility at Rochester, chiefly because it seemed possible that a clever agent could trace back along the capsule's route and identify both the facility and the fact that film was the payload. But a succession of package transfers, arrangements for which were made immediately before the first launch, eliminated most worry.[75]

Apart from such precautions, which were not at all unusual in other satellite reconnaissance programs and which were routine in the case of Corona, the Gambit operation involved one deception that was unique. Colonels Ruebel and Pietz, Major David Bradburn, and Lieutenant Colonel Ralph J. Ford were responsible for an elaborate scheme of misdirection that would have qualified for a good spy novel. It revolved around a wooden box of irregular shape which measured about 12 by 3 by 3 feet. The box—which actually contained three 50-gallon drums filled with water and a set of storage batteries—included several hatches and inspection plates, each secured by heavy hasps and combination padlocks. Under one of the plates was an impressive instrument panel, complete with blinking lights and flickering voltage gauges. It was a very imposing Quaker gun.

The container, painted white and prominently marked with the well-known corporate symbol of Space Technology Laboratories, was for official purposes designated an "Environmental Shipping Container"—ESC for the purpose of what had to seem routine government correspondence.*

The ESC was the basic subject of an elaborate set of pre-written messages to be exchanged between SSD, the launch base at Vandenberg, and several contractor establishments. Designed to provide "tangible evidence" that some government agency other than the Air Force was participating in the "206 Program," the messages were also intended to promote confusion and uncertainty among "those unwitting and peripheral people" who were aware of or involved in the 206 activity. A reasonably astute reader with access to a few of the messages could not but conclude that the 206 payload was something delivered to and maintained by STL immediately before a scheduled launch. By all appearances, each such payload was delivered to the launch site by truck and there installed in the flight vehicle for launch into orbit.

Because of the severe restrictions imposed on the pre-launch handling of the payload (ran the legend), no payload checkout was permitted at the

---

* To my mind, the "ESC" nomenclature was the stroke of brilliance that made this preposterously funny bluff so effective. No self respecting administrator in all of the Air Force could avoid believing in something so classically bureaucratic in title. R.P.

launch site other than normal countdown checks during pre-launch. Too perfect security would negate the desired impression, so a policy of "calculated ineptitude" was adopted which, it was hoped, would expose knowledge of the "ESC" payload to as large an audience as possible while still retaining credibility.

Sixteen messages were involved in the general deception package. One called for a meeting between launch base people and the "payload responsible organization" at SSD to discuss payload problems. Another cancelled the meeting and called for destruction of the first message. A third, from Vandenberg, protested against the lack of a pre-launch checkout and asked for a definition of "ESC." The reply confirmed that no pre-launch payload check would be permitted and called for the assignment of a representative of the test wing who would be instructed in the "operation of the ESC." Others raised questions concerning the minimum pre-launch delivery time needs of the contractor, the test wing, and the "other government agency." The general tenor was that something enclosed in the "environmental shipping container" had to be closely monitored, carefully watched, and handled very, very gently.

On 3 July (eight days before the scheduled launch) an unmarked truck entered Vandenberg. The driver, who exhibited considerable uncertainty about his final destination, eventually was routed to the Lockheed receiving area where a "puzzled" Air Force officer opened one of the shackled hatches of the white-painted crate, briefly disclosed to several nearby personnel an array of blinking lights and pulsing dials, and hastily directed the driver to the General Electric receiving area. There were subsequent meetings between 206 people and the local security officers over the "lapse" in security, all designed to encourage speculation about the peculiar nature of the ESC's contents. At one point the deception proved so successful that it evoked an unplanned inquiry from the Strategic Air Command—which wanted assurances that the "payload" did not contain any nuclear devices to be placed in orbit without due regard for various national laws and international agreements.[76]

In the aftermath of the July 1963 Gambit launch—and those which followed it that year—it became clear that the extremely tight security which had come to surround all military space launches tended to limit the effect of a very ingenious deception scheme. In theory at least, the purpose of such a deception was to convince Soviet intelligence that the 206 payload was not a camera. The ESC maneuver could have that effect only if one of the witnesses to the transaction happened to be an agent or if some of the witnesses mentioned the event and its implications to an agent. By extension, an American agent serving as part of a Soviet espionage net could use the event or certain of the classified messages exchanged during preliminaries as evidence that something presumably radioactive had been orbited in the 206 vehicle. It was conceivable that a Russian agent could be let into the periphery of the security process and fed just enough selected information to mislead him. But nothing of the sort seems to have happened. Instead, a handful of people who had been forcefully indoctrinated with security consciousness were exposed to the ESC deception and were so impressed with the sensitivity of the information they had "accidentally" acquired that they would discuss it only with similarly cleared fellow witnesses. As for the outside world, the only information generally available would indicate that another Atlas-Agena had been launched. The presumption that an Atlas-Agena launched from the vicinity of Vandenberg carried a "Samos" payload was common to journalists and to interested members of the aerospace disciplines. Whether the ESC prank succeeded and was useful was something only the CIA could judge, and the CIA did not say.

Los Angeles was experiencing one of its recurring sieges of "unusual" warm weather during the early afternoon of 12 July 1963. From General Greer's office on the fourth floor of the six-story building that housed the Air Force space program organization, the mountains around the basin seemed no more than slightly solid chunks of the prevailing smog. It was an uncommonly quiet Friday afternoon. Most of the fourth floor offices were empty of officers and senior civilians. General Greer's secretary, and Colonel King's, turned away visitors and telephone inquiries with politely vague phrases: "the general is on TDY, " or "the colonel is out of the complex this afternoon." The double doors at the east end of the main hallway on the fourth floor were closed and latched, which was somewhat unusual, if not unprecedented. Behind them, in a small conference room equipped with a speaker system, project people who could not get to Vandenberg listened intently to the piped-in verbal traffic of launch controllers on the site.

Shortly after two o'clock the gathering broke up. Lieutenant Colonel Ralph Ford, responsible for much of the "ECS" deception that shrouded the purpose of the first Gambit launch from observers, opened the electrically secured door to his office to resume his afternoon routine and found an occasional visitor punching vigorously at his secretary's typewriter. The visitor, a sometime historian, asked in a carefully casual voice, "How'd it go?"

He got a quick grin and the answer, "It's off. It looks good. Real good."

At that moment, 22 months and 17 days after the National Security Council decision to proceed with development of a "covert" alternative to Samos, a new phase in satellite reconnaissance was beginning. The first Gambit had lifted into orbit at 1344 hours, Pacific Daylight Time, on 12 July 1963.

Many of the Gambit program office people had managed to get to Vandenberg to watch the launch. General Greer and Colonel King were at the Satellite Control Facility, at Sunnyvale, watching the launch on remote television and listening to the countdown. Earlier in the day the launch crew had notified Greer that during the final checkout they had uncovered a fault in the Atlas booster that would either force delay or cause reliance on a component not tested to the extent required by specifications. Shouldering aside the oppressive memory of unbroken failures and "partial successes" in the E-5 and E-6 programs, Greer ordered continuation of the countdown. It was a personal decision, taken without consultation with others, based as much on instinct as on the confidence of a program director who had done all that could be done to insure success. For an instant during the launch itself, most observers experienced the horrified conviction that the decision had been wrong, that disaster had come again to the Air Force satellite reconnaissance program. The splashing rocket exhaust of the Atlas knocked out all electrical connections to telemetry and cameras, giving the impression of a major launch stand explosion to observers at Sunnyvale and El Segundo. But seconds later the signals began to come through again, and they said that the Atlas was climbing stolidly toward its selected launch window.

Climbout, separation, and orbital injection occurred as planned. Then for 90 minutes the tense group in the control center had to wait until the satellite completed its first orbital pass and the computers could report precise ephemeris and attitude data. Only then could there be complete assurance that the first Gambit was actually in its intended orbit and that the delicate and complex stabilization equipment was performing its assignment. And after that, another five orbits before the Gambit payload, that complex of optics, electronics, and mechanical devices conceived more than three years earlier, came to life. And then another nine passes before a recovery attempt could be made, and still another wait for information that the capsule had re-entered, had survived its passage through the upper atmosphere, had been arrested in its descent by parachutes, and had been recovered. Even if all possible combinations of failure during orbit, re-entry, and recovery stages were avoided, there remained the ultimate uncertainty: what about pictures?

Both the Atlas and the Agena operated normally, apogee being 116 nautical miles and perigee 107. When excess propellants were dumped from the Agena the reactive force caused an unprogrammed series of vehicle motions that used up considerable portions of the Agena's control gas supply, but enough remained for Agena stabilization during nine orbits.

On the fifth orbital revolution, command controllers turned on the camera for eight strip exposures of 20 seconds each, commanding an identical maneuver on each of the next two orbits. On orbits eight and nine, two stereo pairs and five 20-second strips were exposed—after which the premature exhaustion of Agena stabilization gas forced discontinuance of camera operations.

With the depletion of Agena control gas, the Lifeboat became the only means of recovering the film capsule. The Gambit-Agena coasted through eight uncontrolled orbits after stabilization gas finally was exhausted during orbit nine, ground control activated the "Lifeboat" circuitry during the 17th pass, and on orbit 18 an "execute" signal from the ground station went to "Lifeboat." Routine separation and recovery followed. There was no drama. And nobody minded.

After the reentry capsule had been safely retrieved by C-119 aircraft circling near Hawaii, the Orbital Control Vehicle was separated from the Agena for "solo" tests of various operations.

It maintained stability through orbit 25 and was successfully restarted again on orbit 34, after a period of inactivation. Thereafter spurious commands caused instability. Greer's expectation of the unpredictable had proved reasonable: not only had the orbital control vehicle's mistrusted stabilization system been affected by spurious commands, but the performance of the "reliable" Agena had been unexpectedly degraded when engine propellants were routinely dumped. In the end, the successful operations of "Lifeboat" and "BUSS" (the backup stabilization system) had been essential to the success of the initial mission.

Evaluation of the recovered film indicated an out of focus condition apparently caused by uncompensated temperature changes that affected the face of the primary mirror and by faulty image motion compensation settings. Nevertheless, best resolution on the 74 exposed frames (and nine stereo pairs) was on the order of 3.5 feet; 5-foot ground resolution occurred on several stretches of the 198 feet of exposed film, and average resolution was about 10 feet. With all, it was the best photographic return ever obtained from a reconnaissance satellite, "best" resolution being better than anything previously obtained by a factor of four to five.[77]

Endnotes

1. Notes by A.H. Katz, Rand Corp, 7 Jul 60, on meeting with STL representatives, in Rand (Katz) files.

2. TWX TEC 648/002, K. G. MacLish, EK, to BMD, 20 Dec 60; notes by BGen R. E. Greer, Dir/SAFSP, attached, in SP-3 "G" files.

3. Rpt, "Presentation on G," June 61, prep by SAFSP for SAFUS, in SAFSS files, Gambit; interview, Maj Gen R. E. Greer, Dir/SP, by R. L. Perry, 23 Jan 64.

4. TWX AFDSD-MS-83304, DCS/D USAF to Hq AFSC, 5 Jul 61, in SP-3 files, "G"; memo, J. V. Charyk, SAFUS, to C/S USAF, 8 Aug 61, subj: Atlas Boosters for Support of Space Projects, in SAFSS files, Gambit; Charyk also sent a 23 Jun 61 memo to the C/S ordering the Agena procurement but it did not appear to have found its way to the Gambit files.

5. TWX SAFMS-M-1 163, SAF-MS to SAFSP (Maj John Pietz), 11 Aug 61; memo, Maj H. Howard, SAF-MS, to Maj J. Sides, 10 Aug 61, no subj; TWX SAFMS-M-1 162, SAFMS to Pietz, 11 Aug 61; TWX SSZA 11-8-349, SSD to AFSC, 12 Aug 61; ltr, Maj Gen J. R. Holzapple, Asst DCS/S&L, to SAF-RD, 21 Aug 61, subj: Boosters for Support of Space Projects.

6. TWX AFCVC 64852, Gen F. A. Smith, VC/S USAF, to Gen B.A. Schriever, Cmdr, AFSC, 25 Sep 61, in SAFSS files, Gambit.

7. TWX SCGN 26-9-50, Col R. Nudenberg, Dir/Space Progms, AFSC, to SSD, 26 Sep 61; TWXs SAFSP-F-19-9-1114, SAFSP to SAF-MS, 19 Sep 61, SAFMS-M-1 172 SAFMS to SAFSP, 11 Sep 61, all in SP-3 "G" files.

8. Memo, Maj Gen R. E. Greer, Dir/SP, and Maj Gen O. J. Ritland, Cmdr SSD, to all Gambit-cleared personnel, 14 Nov 61, subj: GAMBIT Project, in SP-3 files; memo, Greer to Gambit-cleared personnel, 16 Nov 61, subj: Cue Ball Security Precautions, same file; TSXs SAFMS-M-1198, SAFMS to SAFSP, SAFSP-F-15-11-143, SAFSP to SAFMS, and SAFMS-M-1197, SAFMS to SAFSP, all 15 Nov 61, all in SAFSS files, Gambit.

9. Memo, J. V. Charyk, SAFUS, to DCS/S&L, 24 Nov 61, no subj; ltr, BGen M.G. Smith, Dir/Sys Services, DCS/S&L, to AF Compt, 27 Nov 61, subj: Project Cue Ball; TWX AFSSV-82581, DCS/S&L, USAF, to AFSC, 27 Nov 61; ltr, Dir/Proc, DCS/S&L, to Hq AFSC, 13 Dec 61, subj: Program Adjustment AF-62-62-600, with MFR on SAFSS copy, initialed with no date but apparently written on 14 Dec, detailing contacts with the budget people; TWX SCGN-22-1-46, AFSC to DCS/S&L, 31 Jan 62; all in SP-3 files, "G".

10. Memo, Maj J. Sides, SAFMS, to BGen R. D. Curtin, 19 Jan 62, subj: Cue Ball; TWX SAFSP-F-1-2-190, Maj Gen R. E. Greer to Curtin, 1 Feb 62, both in SP-3 files; memo, LtGen James Ferguson, DCS/S&L, to SAFUS, 22 Jan 62, subj: Cue Ball, in SAFSS files, Gambit.

11. TWX SAFMS-INS-M-2017, BGen R. D. Curtin to Maj Gen R. E. Greer, 16 Feb 62, in SP-3 files, Funding; ltr, Col Ch, R&D Programs Div, DCS/S&L, to AFSC, 13 Feb 62, SAFSS files, Gambit; MFR, Maj J. Sides, SAFMS, 7 Mar 62, subj: Trip Report, SP, 28 Feb 62, Gambit, in SAFSS files.

12. MFR, Col J. L. Martin, SAFMS, 13 Feb 62, subj: SAFUS SAFSP West Coast Conference 9 Feb 62, in Martin's files, SAFSS; MFR, Sides, 7 Mar 62.

13. MFRs, BGen R. E. Greer: 6 Jan 61, subj: Trip 4-5 Jan 61, and 16 Jan 61, Time and Materials Contract EKC, both in SP-3 Gambit files.

14. Memo, Maj Gen R. E. Greer, Dir/SP, to J. V. Charyk, SAFUS, 10 Jul 61, no subj, in SAFSS files, "G".

15. TWX SAFSP-F-18-9-10, SAFSP to SAFMS (Maj H. Howard), 18 Sep 61; TWX SAFMS-M-1-187, BGen R. D. Curtin, SAFMS to Maj Gen R. E. Greer, SAFSP, 20 Oct 61, both in SAFSS files, "G"; TWX SAFSP-F-12-8-99, SAFSP to EK, 23 Aug 61, in SP-3 files, Gambit.

16. Briefing charts, 9 Nov 61, "Project Cue Ball," in SP-3 files.

17. MFR, Maj J. Sides, SAFMS, 30 Jan 62, subj: Trip Report- G (22-26 January 1962), in SAFSS Gambit files; ltr, Maj Gen R. E. Greer, Dir/SP, to SAFUS, 28 Feb 62, subj: Action Pursuant to 9 February 1962. In SP-3 files.

18. MFR, Sides, 30 Jan 62; TWX SAFMS-DIR-62-25, BGen R. D. Curtin, SAFMS, to Maj Gen R. E. Greer, Dir/SP, 1 Feb 62, in SP files, General.

19. MFR, Col J. L. Martin, D/Dir SAFSS, 8 Mar 62, subj: SAFUS-SAFSP EK-GE Visit on 5-6 Mar 62, in SAFSS files, Gambit.

20. Ltr, Greer to SAFUS, 28 Feb 62.

21. TWX SAFSS-DIR-M-2024, SAFSS to SAFSP, 19 Mar 62, in SAFSS files, Gambit.

22. Memo, Maj Gen R. E. Greer, Dir/SP, to J. V. Charyk, SAFUS, 14 Mar 62, subj: 698AL Program Enhancement, in SAFSS files, Gambit corres.

23. TWX SAFSS-DEP-M-2023, SAFSS to Maj Gen R. E. Greer, Dir/SP, 19 Mar 62, in SP-3 files, "G".

24. Memo, Maj Gen R. E. Greer, Dir/SP, to J. V. Charyk, SAFUS, about 15 Mar 62, subj: Contractor "X" in SAFSS files, Gambit; TWX SAFSS-DIR-M-2024, 19 Mar 62; memo, Greer to Charyk, 14 Mar 62.

25. TWXs: SAFSP-F-10-4-299, SAFSP to SAFSS, 10 Apr 62; SAFMS-INS-M-2035, SAFSS to SAFSP, 20 Apr 62; SAFSS-PRO-M-2038, SAFSS to SAFSP, 20 Apr 62; SAFSP-F-28-4-239, SAFSP to SAFSS, 30 Apr 62, all in SP-3 Funding files; MFR, Martin 8 Mar 62.

26. TWXs: SAFSP-F-4-5-243, SAFSP to SAFSS, 4 May 62; SAFSP-F-10-247, Maj Gen R. E. Greer, Dir/SP, to J. V. Charyk, SAFUS, 10 May 62; SAFSP-F-10-5-248, Greer to BGen R. D. Curtin, Ofc Space Sys, 10 May 62; SAFSS-DLRM-2041, SAFSS to SAFSP, 14 May 62; SAFSP-F-15-5-255, Col Q.A. Riepe to Greer (in SAFSS), 15 May 62; SAFSPF-17-5-257, SAFSP to SAFUS, 17 May 62, all in SP-3 files, "G"; memo, T. D. Morris, Asst SOD/Instal and Log, to SAFUS, 31 May 62, subj: Industrial Facilities Expansion, Project 698AL, in SAFSS files, Gambit.

27. Memo, L. C. Meyer, Ch, Mis and Space Sys Div, Dir/Budget, AsstSAF (Fin), to J. V. Charyk, SAFUS, 17 Apr 62, no subj, in SP-3 files, Funding; TWX SAFSP-F-17-4-232, Maj Gen R. E. Greer, Dir/SP, to BGen R. D. Curtin, Ofc Space Sys, SAF, 17 Apr 62, in SP-3 files, "G".

28. Memo, W. F. Sampson, Aerospace Corp, to Col Q. A. Riepe, Dir/206 Progm, 19 Jan 62, subj: Air Retrieval for Program 483A, in King files; Interview, Maj Gen R. E. Greer, Dir/SP, by R.L. Perry, 11 Sep 64.

29. Greer interview, 11 Sep 64; msg, Maj Gen R. E. Greer to J. V. Charyk, SAFUS, and Col J. L. Martin, Dir/NRO Staff, approx 3 Aug 62, in SAFSS files, 162-Disc.

30. TWX, SAFSS-DIR-M-2072 and 2073, both J. V. Charyk, SAFUS, to Maj Gen R. E. Greer, Dir/SP, 28 Jul 62, in SP-3 files.

31. Msg, approx 3 Aug 62.

32. Memo, to Col Q.A. Riepe, Dir/Prog 698AL, 1 Aug 62, Sub.): Pro-Con-Overwater vs Land Recovery, in SP-3 files, Land Recov; the affair of the station is covered in MFR, Maj J. Sides, SAFSS, 7 Mar 62, subj: Trip Report, SP, 28 Feb 62, Gambit, in SAFSS files, Gambit, in TWX Dean Rusk, Secy/State, to American Embassy, 2 Jul 62, cy in SAFSS Gambit files, and in various TWXs in those files.

33. Study, "Pros and Cons of Satellite Recovery Methods," Maj Gen R. E. Greer, Dir/SP, 4 Aug 62, in SP-3 files,Land Recov.

34. Memo, prep by Capt F.B. Gorman (USN), SAFSP Plans Oft, 17 Aug 62, subj: Program Recovery Information Desired by Dr. Charyk; TWX SAFSP-F-14-8-321, Maj Gen R. E. Greer, Dir/SP, to Col J. L. Martin, Dir/NRO Staff, 14 Aug 62, both in SP-3 files, Land Recov; TWX AFSSA-AS-1-7768S, DCS/S&L, USAF, to Cmdr AFSC, 24 Aug 62, in SAFSS files, Gambit.

35. TWX SAFSS-DIR-M-2082, SAFSS to Dir/SP, 24 Aug 62, inSAFSS files, Gambit.

36. TWX SAFSS-DIR-M-2082, SAFSS to Dir/SP, 24 Aug 62; msg, CLA. to SAFSP, 18 Sep 62, in (Leach) files; TWX SAFSP-F-7-9-333, Col R.A. Berg, Asst Dir/SP, to Col J. L. Martin, Dir/NRO Staff, 7 Sep 62, in SP-3 files, "G", summarized the course of the earlier .station negotiations and the uncertainty as to their status. TWX SAFSP-F-21-9-343, Maj Gen R. E. Greer, Dir/SP, to Martin, 20 Sep 62, confirms agreements of the 18 Sep 62 meeting, details of which are contained in memo for the record, prep by 19 Sep 62, in Col King's files; TWX SAFSP-F-20-2-530, SAFSP to SAFSS, 20 Feb 63, concerns the Annette Island station and the doppler radar requirements.

37. Msg, , CIA to SAFSP, 18 Sep 62; TWX SAFSPF-21-9-343, Maj Gen R. E. Greer, Dir/SP, to Col J. L. Martin, SAFSS, 20 Sep 62, in SP-3 files, Finance; msg, SAFSP to CIA, 24 Sep 62, and msg, CIA to SAFSP, 28 Sep 62, both in SAFSS files, Gambit.

38. Rpt, Gambit Report for FIAB, 28 Sep 62, in SAFSS files, Gambit.

39. Greer interview, 11 Sep 64.

40. TWX SAFSS-DIR-M-2095, J. V. Charyk, SAFUS, to Maj Gen R. E. Greer, Dir/SP, 3 Oct 62, in SAFSS files, Gambit.

41. Interview, BGen J. L. Martin, D/Dir/SP, by R. L. Perry, 18 Sep 64; interview Maj Gen R. E. Greer, Dir/SP, by Perry, 22 Nov 63.

42. TWX SAFSP-F-5-10-364, Maj Gen R. E. Greer to members of the 206 Progm Ofc, 30 Oct 62, subj: New Assignments, in SP-3 files, Progms.

43. *Endnote in text but missing from endnote list.*

44. SSD Spec Order, 28 Nov 62, in SSD Mil Pers Div files.

45. Greer interview, 22 Nov 62.

46. Interviews: LtCol J. Pietz, 14 Sep and 7 Oct 64; Col W. G. King, 7 Oct 64; Maj Gen R. E. Greer, 15 Sep 64, all by R. L. Perry.

47. Memo, Maj Gen R. E. Greer, Dir/SP, to Col W. G. King, D/Dir/206 Progm, 7 Dec 62, subj: H-30 Capsule, and 1st Ind, King to Greer, 10 Dec 62, in King's files.

48. Memo (1st Ind), King to Greer, 10 Dec 62; King interview, 7 Oct 64.

49. TWX SAFSP-F-8-11-409, Maj Gen R. E. Greer, Dir/SP, to J. V. Charyk, SAFUS, 8 Nov 62, in SP-3 files, Programs; rpt, Security of Satellite Reconnaissance Activities, 25 Jun 62, in SAFSS files, Basic Policy; interview, Greer by R. L. Perry, 23 Jan 64.

50. TWX SAFSS-1-M-2119, SAFSS to SAFSP, 20 Nov 62, in SAFSS files, Gambit.

51. TWX SAFSP-F-30-11-432, Maj Gen R. E. Greer, Dir/SP, to Col J. L. Martin, SAFSS, 30 Nov 62; TWX SAFSS-1-M-2129, Martin to Greer, 30 Nov 62, both in SAFSS files, Gambit; interview, LtCol John Pietz, SP-3, 6 Jun 63, by R. L. Perry; Greer interview, 22 Nov 62.

52. TWX SAFSS-1-62-178, SAFSS to SAFSP, 11 Dec 62; interview, LtCol J. Pietz, SP-3, by R. L. Perry, 22 Jul 64; TWX SAFSS-1-M-2138, SAFSS to SAFSP, 19 Dec 62 (confirming the hitchup, roll-joint, lifeboat, and H-30 changes), both in SAFSS files, Gambit; TWX, SAFSP-F-24-1-491, Maj Gen R. E. Greer, Dir/SP, to NRO Compt, 24 Jan 63, (summarizing "white" costs), in SP-3 files, Funding; msg, SAFSP to 9 Jan 63 (concerning "black" costs), in (Leach) files.

53. Greer interview, 22 Nov 64, 11 Sep 64; Pietz interview, 6 Jun 63.

54. Pietz interview, 6 Jun 63; Greer interview, 22 Nov 63; TWX S A FSS -1- M-2138, 19 Dec 62, msg, Maj Gen R. E. Greer, Dir/SP, to J. V. Charyk, SAFUS, 6 Dec 62, in SAFSS files, Gambit.

55. Rpt, Program 206 Detail Schedule, Oct 62, in SP-3 files, Progms; rpt, Program 206 Bi-weekly Activities, Rpt, 2 Dec-22 Dec 62, in SP-3 files, Gambit.

56. Memo, Col W. G. King, D/Dir 206 Progm to Maj Gen R. E. Greer, Dir/SP, 21 Dec 62, subj: Horizon Sensors; memo, Greer to King, 8 Jan 63; memo, King to Greer, 11 Aug 64, subj: Improved Horizon Sensors; ltr, King to GE/ASPD, 17 Sep 64, same subj; all in King's files.

57. Memo, Col W.G. King, D/Dir/206 Progm, to Maj Gen R. E. Greer, Dir/SP, 7 Jan 63, subj: 206 Program Funding; memo, King to Greer, 15 Jan 63, subj: Review of Changes to Program; memo, Greer to King, subj: 206 Program Funding, all in King's files.

58. Memo, Col W. G. King, D/Dir/206 Progm, too'" 22 Jan 63, subj: Estimate of Funding Required to Provide Program Changes Directed by SAFUS, 19 Dec 62, in King's files.

59. Memo, Col W. G. King, D/Dir/206 Progm, to Maj Gen R. E. Greer, Dir/SP, 27 Feb 63, subj: 206 Costs and Current Contract Negotiations, in King's files.

60. Memo, LtCol J. Sides, SAFSS, to Col J. L. Martin, Dir/NRO Staff, 31 Oct 62, subj: "G" Camera, in SAFSS files, Gambit.

61. Memo, A. C. Lundahl, Dir/NPIC, to Dir/NRO, 24 Jan 63 (with undated longhand note by H. Scoville, Dir/Res CIA), subj: Specific Technical Requirements Satellite Photographic Systems; TWX SAFSS-6-M-0051, SAFSS to SAFSP, 28 Feb 63; memo, Col J. L. Martin, Dir/NRO Staff, to Dir/NPIC, 2. May 63, subj: Stellar/Index Camera for GAMBIT, all in SAFSS files, Gambit; TWX SAFSP-F-18-2-526, SAFSP to SAFSS, 18 Feb 63, in SP-3 files, "G".

62. TWX SAFSP-F-5-3-548, SAFSP to SAFSS, 5 Mar 63; TWX SAFSS-1-M-0070, SAFSS to SAFSP, 28 Mar 63; TWX SAFSPF-16-4-641, SAFSP to SAFSS, 16 Apr 63, all in SAFSS files,Gambit.

63. TWX SAFSS-6-M-0059, SAFSS to SAFSP, 12 Mar 63; TWX SAFSP-F-14-3-579, SAFSP to SAFSS, 15 Mar 63; TWX SAFSS-1-M-0070, 27 Mar 63; TWX SAFSS-1-M-0071, Col J. L. Martin, Dir/NRO Staff, to Maj Gen R. E. Greer, Dir/SP, 28 Mar 63, TWX SAFSS-6-M-0074, Martin to Greer, 2 Apr 63, all in SAFSS files, Gambit.

64. TWX SAFSS-1-M-0105, SAFSS to Maj Gen R. E. Greer, Dir/SP, 10 May 63; memo, B. McMillan, D/NRO, to Greer, 10 May 63, no subj; memo, Greer to SAFUS, 15 May 63, subj: Budgeting and Procurement, all in SP-3 files, Funding.

65. Memo, Col W. G. King to Maj Gen R. E. Greer, Dir/SP, 10 May 63, subj: Preliminary Feasibility Report - Dual Recovery Capability for GAMBIT; indorsement, Greer to Col J. L. Martin, SAFSS, 13 Jun 63, same subj, both in SAFSS files, Gambit.

66. TWX SAFSP-F-11-2-517, SAFSP to SAFSS, 11 Feb 63; TWX SAFSP-F-13-3-575, SAFSP to SAFSS, 13 Mar 63; TWX SAFSP-F-27-3-608, SAFSP to SAFSS, 27 Mar 63; TWX SAFSP-F-2-4-619, SAFSP to SAFSS, 2 Apr 63; TWX SAFSP-F-16-4-642, SAFSP to SAFSS, 16 Apr 63, all in SAFSS files, Gambit.

67. TWX SAFSP-F-16-5-706, Col W.G. King to LtCol J. Sides, SAFSS, 17 May 63; TWX SAFSP-DIR-15-5-1, SAFSP to SAFSS, 15 May 63; TWX SAFSS-1-M-0119, SAFSS to SAFSP, 23 May 63; TWX SAFSP-F-6-737, King to Col J. L. Martin, SAFSS, 5 Jun 63, TWX SAFSS-1-M-0134, 6 Jun 63, all in SAFSS files, Gambit.

68. TWX SAFSS-1-M-0109, SAFSS to SAFSP, 15 May 63; memo, Maj Gen R. E. Greer, Dir/SP, to Maj Gen B. I. Funk, Cmdr, SSD, 15 May 63, subj: 206 Program; memo, Funk to Greer, 31 May 63, same subj; TWX, SAFSP-F-17-6-758, Greer to B. McMillan, SAFUS, 17 Jun 63, all in SP-3 files, "G".

69. TWX SAFSP-F-22-5-718, Col W. G. King, SAFSP, to LtCol J. Sides, SAFSS, 22 May 63, in SAFSS files, Gambit.

70. Msgs, CLA to D/NRO, 4 Jan 63; and CIA to SAFSP, 18 Jan 63, both in (Leach) files.

71. TWX SAFSP-F-19-1-498, SAFSP to 6594th ATW, 29 Jan 63, in SP-3 files, G.

72. Interview, Col J. W. Ruebel, SP-3, by R. L. Perry, 29 May 63.

73. Interview, Col R.J. Ford, SP-3, by R. L. Perry, 17 Sep 64.

74. Interview, LtCol J. Pietz, SP-3, by R. L. Perry, 29 May 63.

75. Msgs, EK to CIA, 11 Jul 63; and D/NRO to SAFSP, 1 Jul 63, both in Leach files, TWX SAFSP- 8-21-588, SAFSP to SAFSS, 21 Mar 63, in SAFSS files, Gambit.

76. "Book," Gambit Payload Deception Operation, May 63, prep by LtCols R. J. Ford and J. Pietz, SP-3, in files; ltr, J. D. Hansford, GE, to Col R. J. Ford, SAFSP, 21 Oct 63, subj: Memo Report on STL Payloads, 951, 952 and 953 in SP-3 files; Ford interview, 14 Sep 64.

77. Author's notes, 12 Jul 63; rpt, Gambit Program Summary Report 1960-1967, prep by R. Perry, Sep 67 (hereafter cited as G Summary); rpt, Summary Analysis of Program 206 (GAMBIT), prep by 29 Aug 67 (hereafter cited as 206 Summary).

GAMBIT–1 OPERATIONS

The flight portion of the Gambit* Project offered interesting departures from the "normal" cycle of research, development and operations observed by most DoD development agencies. It owed much in that respect to the precedent of the Corona program, the only earlier satellite reconnaissance activity that could even casually be called successful. Although nominally divided between development and operational phases, the dividing line coming after the fifth flight, the de facto value of each flight was measured in various ways. Criteria ranged from the performance of the Gambit vehicle and camera system through the success of experiments and equipment modifications which were of development significance to a variety of projects, including Gambit-3. As for operational value, data of appreciable intelligence worth were collected as early as the fifth flight, and the quantity of such data continued to increase at a steady rate throughout the life of the original Gambit program.

While technical developments including the refinement of hardware and the introduction of new manufacturing techniques were of obvious significance, other and perhaps less tangible aspects of the Gambit project had greater potential long-term value. They were mostly of a program management sort. They included such areas as security devices necessary to "black" programs and management techniques for ushering a program through flight test into operations. General Greer's stubborn insistence that "one good picture" was the only valid goal of the first flight lost much of its dramatic impact once success became the norm rather than the exception for new programs. But it was almost certainly one of the pivotal reasons for the early success of the Gambit project.[1] Other projects in the space reconnaissance program had fallen almost entirely from the weight of overly ambitious early flight objectives. The result, with uncommon regularity, had been catastrophic failure and consequent abandonment of the program. Whatever had been invested was lost. Greer's forte had been that of a midwife to the new project—overseeing and guaranteeing a successful birth and infancy. His successor, Brigadier General John L. Martin, Jr.,** proved to be particularly adept at raising the child to maturity. Martin's handling of a midstream crisis (three successive catastrophic failures midway through the flight program) by re-orienting contractual incentives served as a model for future contracting practices as well as solving the problem of the moment. The elements of the incentive program were probably of less importance than its conceptual basis. It represented an acknowledgement that the goals of a project changed as it outgrew its developmental constraints, and that incentives suitable for one phase were not necessarily appropriate to another.

Less than two weeks after the first Gambit flight aimed at "one good picture," General Greer advised the project director, Colonel W. G. King,*** that he very much wanted "two in a row."[2] That was to become the watchword for the second Gambit flight. While Greer was gratified by the success of the first flight, he appreciated that unwisely ambitious second-flight objectives could damage the program just as much as an unsuccessful first flight. Paradoxically, the very success of the first flight raised expectations for subsequent flights and could be expected to make later failures even more unsettling to those who ultimately controlled project funding. If enough success could be tucked away in the flight history of the basic hardware, then downstream failures could be treated as local problems rather than indications of a flawed conception. While no one knew how many successful flights or how much good output was required to create this aura, King and Greer were both quite positive that at least the second flight would have to be a pronounceable success.

The operational plan for the second flight called for three days of flight in a hitch-up mode. During the first flight control over the orbital control vehicle (OCV) had been lost after seven orbits, which did not represent enough of a test to justify making the success of the second flight dependent on its proper functioning. King decided that the Agena should be relied on once again for orbital control during that portion of the flight when primary mission objectives were to be satisfied. Those objectives were the demonstration of best resolution from the camera and successful recovery of the film. The secondary mission goal was to demonstrate controlled independent flight by the

---

* On nomenclature: Gambit operated under a considerable variety of titles and numerical designators. Two principal designators are used hereafter in this manuscript. The name Gambit ordinarily is used to identify the vehicles, and the program, that included any version of the original 77-inch (focal length) camera and the original single-recovery-vehicle film retrieval system. Gambit-3 (for "Gambit-Cubed") is the comparable designator for the systems and program that involved the 160-inch lens and the "double bucket" recovery system. Where both the original Gambit and Gambit-3 are being discussed, the term Gambit-1 has been used to lessen confusion. It will be recalled that Gambit carried the "white" titles "Exemplar" and "Cue Ball." In 1961 and 1962, the latter name being associated with the unclassified program number 483A, mostly for accounting purposes. In mid-1962 the terminology "Program 698-AL" was applied as an unclassified identifier for Gambit; it subsequently was changed to "Program 206, "the terminology ordinarily used until the original Gambit gave way to Gambit-3. That follow-on program, which officially became "Gambit" in 1969 (the "G³" or "Gambit-Cubed" description was formally dropped at that time), carried the numerical designator "Program 110."

** Both Greer and Martin retired as Major Generals.
*** Later a Brigadier General and Martin's successor

OCV, but not until after three days had been logged on orbit in the hitch-up mode.[3] A cautionary note was injected by anxiety over the operation of the Agena, which had malfunctioned during the first flight. Greer and King decided that ground controllers should be prepared, beginning with the second revolution, to separate the OCV-RV combination from Agena at the first sign of trouble.[4]

Although master schedules called for one Gambit launch every 40 days, making 6 September nominally six days late,[5] early Gambit flights were acknowledged to be development flights, so neither the schedule nor the slippage was considered critical. What was important was to precede each flight with a full analysis of the failures of the previous flight—and to incorporate corrective features. The first two years of Gambit flights were to be marked by steady increases in pre-flight testing and by the installation of telemetering devices to monitor in-flight failure modes. That trend developed from a gradual understanding that although the proper dictum was to correct each flight's failures before the next, the extent of effort needed to successfully perform that task had initially been underestimated.

Underestimating may have been the least important of several influences. In the early stages of the program, its managers were justifiably worried that it might be cancelled. The record of earlier failure in other satellite reconnaissance efforts, and financial overruns in the Gambit program provided reason enough for that worry. In any case, Greer perceived the urgency of extensive pre-flight tests to enhance the probability of program success even at the cost of schedule slippages. He had gone a long way toward hedging his bet by massive simplification of the Gambit hardware and early flight operations. While lack of adequate test data continued to trouble the program for some months, it was clear in retrospect that Greer made the right tradeoffs. They were clearly responsible for the regular success and smooth progress which marked the program for all but the middle portion of its life.

Another factor of some considerable importance in the perceived vulnerability of the early Gambit program was a fundamental difference of viewpoint between the CIA and the photo-intelligence community, on the one hand, and Greer's organization plus the NRO staff, on the other. Admittedly, the Gambit group saw their mission as one of correctly exposing and efficiently recovering film. They were less concerned with the intelligence content of the product, as such. The intelligence community was preoccupied with the information content of the film; its members were willing to accept the risks of mission failure if the quality and quantity of intelligence returns might be enhanced thereby. Greer's people were not, arguing reasonably enough that failed missions returned nothing of value to anybody.[6]

The Gambit countdown on 6 September was uneventful; launch occurred at 12:30 local time. All went well. Perigee was 102 nautical miles. During fifty-one hours on orbit, the hitched vehicle completed 34 orbits and exposed some 1930 feet of film, some in stereo pairs but the most in single frames. The mission covered ten different intelligence targets. On the 34th revolution, the reentry vehicle was detached and successfully recovered by air catch.[7]

During separation of the Agena from the orbital control vehicle, a malfunction of the pneumatic system caused a rapid loss of stabilization gas. As a result, the major objective of the solo flight of the OCV—operation of the stabilization subsystem—could not be demonstrated. The OCV was deboosted before completing any of its planned 49 revolutions.

Gas leak problems were not confined to the stabilization system. On revolution 31 of the hitched flight, the primary camera door failed to close (during the 20th opening-closing cycle). That too was due to a gas leak, but in the door actuator. That event pressaged a problem which was to recur in one form or another until the pneumatic systems were eventually replaced, in both primary and secondary modes, by electro-mechanical actuators. But that modification was not effective until the 26th Gambit.

Initially, the pneumatic door actuator failure did not appear to be a major problem. In terms of product, the flight was hugely successful. During hitch-up, the cameras provided ground resolution of 2.5 feet. The contractual specifications called for two to three feet, so one of the three major objectives of the project had been satisfied on the second flight. (The other objectives were an operational life of five days and the ability to point the camera at will. A completely successful five-day flight was almost two years and 17 flights away; pointing accuracy was to be demonstrated by the seventh flight.)

The Gambit flight program developed three major classes of problems. The first and least frequent but most persistent appeared as failures of various pneumatic subsystems. Another class of problems included one-time failures, which once corrected did not reappear.

The third class of problem was intellectually the most interesting and operationally the most frustrating. Throughout the program instances of seemingly random failure occurred in components which had functioned correctly for many flights. The problem would persist through three or four flights, notwithstanding strenuous correction efforts, before succumbing. While there was nothing mysterious about the recurrence of a given failure, the sudden appearance of one where none had existed earlier was unusual for space vehicles, used only once and normally immune to wearout as such. No fully satisfactory explanation of the phenomenon ever appeared, although transient quality control and test program faults were generally blamed.

The aftermath of the second flight brought a renewal of controversy about the paramount objectives of early Gambit flights. An analysis of the photographs recovered from the second Gambit showed consistently high quality until the 31st orbit. The failure of the main camera doors to close thereafter with consequent optical problems caused by temperature transients, had caused a softening of image quality and some loss of resolution. But the resolution achieved during the initial portion of the flight was sufficient to distinguish details like aircraft engine nacelles, small vehicles and even maintenance equipment.[8] Thus, for the first time an orbiting camera had returned detail at levels previously obtained only from aircraft. In effect, only three years after aircraft overflights of the Soviet Union had been discontinued, satellite reconnaissance had more than filled the gap. First, Corona had returned coverage of areas most U-2's could not reach or could not safely overfly, and now Gambit had returned detail not greatly inferior to that produced by U-2 cameras. But the Gambit returns had been limited; 1950 feet of film was not a large return, only ten targets having been covered, but more important, the OCV on which eventual routine coverage would depend had not yet functioned properly. Pointing accuracy demonstrations were lacking. Although the Gambit achievement represented remarkable progress and excellent research and development results, it did not yet constitute a basis for good recurring coverage of the Soviet Union. And information of that sort, at resolutions much better than Corona could provide, was an urgent national goal.

Given that the high resolution potential of Gambit had been demonstrated and pointing accuracy using the orbital control vehicle had not, Greer decided that for the third flight the primary objective should be demonstration of the operation of the vehicle. High resolution photography of intelligence targets would be relegated to a secondary objective.

McMillan approved Greer's decision on 17 September, after which King forwarded the formal statement of objectives and priorities to Washington.[9] The target launch date was set for 22 October.

Operations were to be maintained for two days in the hitch-up mode and two additional days in solo flight.

That operational plan prevailed until 24 October. While technical problems that had caused a three-day launch postponement were being resolved, McMillan reversed his earlier position, telling Greer he wanted to "clarify" the objectives of the Gambit program. He explained that the effectiveness of the program could not be judged by the ground area covered or the amount of film exposed successfully. Rather, effectiveness would be judged in terms of the number of high priority targets for which high resolution stereo pairs could be obtained. Primary efforts for following Gambit missions were to concentrate on obtaining the best possible ground resolution over larger numbers of "denied area" targets. Orbital control, as such, was to be a secondary consideration. Further, development-oriented flights were to end as soon as possible.[10] Full operational missions were wanted at the earliest achievable date.* In historical summaries, the primary objective of the third flight is duly recorded as obtaining maximum information from high resolution photography, and the secondary objectives as demonstrating capabilities of the orbital control vehicle and proving the feasibility of a five-day flight.[12]

But post-fact notations do not necessarily reflect the course of real events. The 34 revolutions of the third Gambit vehicle in the hitch-up mode were what had been specified in Greer's message of 13 September, in any case. The only other relevant evidence concerning real objectives of the flight is the amount of film used and the number of targets photographed. One hundred feet less film was recovered from the third Gambit mission than from the second, and the amount actually exposed on orbit (rather than in pre-launch tests) was less than half that of the second flight. Four targets were photographed compared to ten of the second flight.

Greer's principal concern was to demonstrate, systematically, that each essential element of Gambit hardware was functional. He was satisfied that sufficiently high resolution photography could

---

* In September and October 1963, McMillan had the first of several major brushes with Dr. Albert D. Wheelon, who had become the CIA's Deputy Director, Science and Technology, in July. Even earlier, McMillan had been exposed to several pointed suggestions that Greer's organization be instructed to subordinate its R&D orientation to an intelligence-return orientation. Although no directly relevant documents from that period have survived that detail the Gambit flight goals disagreement of July-October, indirect evidence of the pressures on McMillan and of his reluctance to reorient Gambit flight objectives is found in his own memoranda. The instructions to Greer on 24 October reflect an effort to compromise existing differences, requiring—or formally stating—an unwritten agreement between McMillan and Roswell Gilpatric, Deputy Secretary of Defense, to place greater emphasis on quickly ending the R&D phase of the Gambit program.[11]

be obtained. If he had doubts, they were resolved by the results of the second and third missions. His next concern was for the orbital control vehicle. Dr. McMillan had to cope with different constraints. One of his problems was that only the first three Gambit flights had been represented to "high authority" as developmental. While he and Greer were completely agreed on the need for more developmental flights, McMillan also wanted to be able to display intelligence returns that could substantiate claims of operational utility for Gambit. He was not prepared to reverse Greer, in any case. On 12 February 1964, he authorized developmental flights to continue beyond the fifth Gambit. Indeed, he ruled that flights would only be designated operational after "several" four- or five-day flights had been successful.[13] Since the first such mission was not scheduled until August 1964, Greer's arguments clearly had prevailed.

By February 1964, however, it had become almost certain that Gambit would return large quantities of highly valuable photographic intelligence, and in the reasonably near future. McMillan therefore broadened his position in dealing with "users," and supported Greer's well-based convictions both privately and publicly.

Some additional cause for anxiety about the success potential of the third Gambit resulted from the history of the orbital control vehicle. It had originally been used in thermal-vacuum testing of the satellite vehicle. It was subsequently refurbished and assigned the third Gambit flight. By 25 October 1963, all systems had been checked out and, for the first time, propellants were loaded that would support orbital adjust maneuvers. At one minute before noon, local time, the booster was ignited and the third Gambit space vehicle was put into orbit.

Whatever the preliminary uncertainties about mission objectives and equipment, the result was a mission that conformed to flight goals from first to last. Not only was the photography as good as that of the second flight, but the recovery was routine and the orbital vehicle, in solo flight, successfully demonstrated orbital adjust and de-orbit capabilities.[14] During the solo flight the stabilization subsystem was exercised extensively and found to operate as specified. Of the 38 Gambit flights ultimately undertaken (not including Gambit-3) this was one of only three which was perfect in the sense that it was unmarred by any failure, major or minor. The other two such missions came at the very end of the Gambit-1 program.

The general quality of the photography was judged to be "...better and more consistent..." than that of either of the first two missions. Photographs from that mission were the first to show identifiable figures of people on the ground—from a distance of some (The scene was a football field in Great Falls, Montana.) In one picture a place kicker could be seen putting the football in place while other players moved into position. In a second photograph the players had lined up, ready for the kickoff.[15]

Its first three flights having been successful, Gambit secured the virtual guarantee of continued funding Greer had sought. Lanyard, the "insurance" surveillance system that backed up Gambit, was cancelled.[16] The Gambit space vehicle and payload having been proven, the only backup necessary in the future was an operational one, satisfied by the production of Gambit vehicles for continued use.

Although photography obtained from the first three flights of Gambit had little intelligence value, it had superb resolution. Orbital adjustment maneuvers had been carried out successfully, as had solo flight of the orbital control vehicle, after separation of the payload capsule for reentry. The entire system had an evident capability of performing five-day operational missions. The problems thus far encountered appeared to be manageable.

The next step was a mission which fully explored operational capability. The stated primary mission of the fourth Gambit was to obtain high-quality reconnaissance photography, as would be the case with every flight thereafter, to the end of the program. The secondary mission goal was demonstration of five-day longevity.[17]

The fourth flight vehicle differed substantially from the first. The stellar index cameras (being procured through a black CIA contract with Itek) still were not available, but most of the other major features of Gambit, as originally planned, were present. The hitch-up option was dropped, giving the orbital control vehicle its first opportunity to display its capabilities in a "live" test. Such changes had once been planned for the sixth mission, but success in the solo operation of the OCV had been so encouraging that program managers concluded that no more useful information could be generated by further dry runs. A "Lifeboat" system was installed for the first time in the fourth Gambit. (Also known as the back-up stabilization system [BUSS], "Lifeboat" utilized the earth's lines of magnetic force as a reference to stabilize the vehicle in flight. It had its own control gas supplies and command circuitry, separate from the primary stabilization system. Originally conceived as temporary insurance against the failure of the primary system in early missions, BUSS survived to the last in the original series of Gambit flights and was last used to bring back the reentry vehicle of Gambit number 36.)

Launch occurred in the early afternoon of 18 December 1963. Separation of the Agena and the OCV and orbital injection were nominal. Launch controllers were slightly uneasy because of an uncorrected test failure which had occurred during countdown. They had been unable to change the crab position of the primary mirror. The failure was considered random; probably due to a short circuit between the crab servo and the command decoder relay box. In any case, no corrective action was considered necessary. In any event, the short circuit persisted, but the anticipation of "no problem" proved accurate.

The first four orbits were normal. During the fifth revolution, however, the satellite vehicle began to tumble and all orbital control gas was expended in efforts to stabilize it. The precipitating failure was in a small heat controlling device responsible for maintaining acceptable temperatures around the rate gyro. The heater operated at full output the first four orbits. The result was a heat-induced malfunction of the rate gyro, causing massive instability of the space vehicle. Mission controllers decided to deboost and recover the RV during the 18th revolution.

Because the primary stabilization gas supply had been fully exhausted at the end of the fifth revolution, controllers had to call on the Lifeboat system. Tumbling was so extreme that BUSS could not fully suppress it during the next 13 revolutions. By the time the recovery was attempted, BUSS gas was all but depleted. The result was that deorbit of the RV was inaccurate and reentry occurred 720 miles downrange from the planned location. Nevertheless, because they had been warned in advance of that likelihood, the recovery team was still able to recover the capsule by air catch.

Of the missions assigned to the fourth flight, only one was demonstrated, and that only in part: the BUSS, newly installed on this flight, performed very well in an unexpected emergency. It had allowed the recovery of the capsule when all else would have been unavailing.

Although the fourth Gambit had experienced catastrophic mission failure and had performed none of its primary assignments, the fault was so localized that it caused no significant change in the system. The only alteration suggested by the failure was to wire a back-up heater switch in the rate gyro assembly to turn off the heater when a critical temperature was reached.[18]

Even though the early Gambit flights were developmental and telemetry of error mode information was essential, much instrumentation essential to detailed system operation monitoring was not incorporated in the first four Gambit vehicles. Limited mission objectives in combination with budget problems partly explained the omission. After the first three flights, that rationale evaporated, to be replaced by another: program success. Hindsight suggested that instrumentation was inadequate to the needs of development.

For various reasons, the launch schedule for the fifth Gambit was allowed to slip to the extent of about a month. During the interval between missions four and five, McMillan accepted the premise that ten development flights should be programmed before operational status could be claimed. Gambit was not to be considered operational, McMillan concluded, until, "...several completely successful four- or five-day missions have been accomplished and all significant operational limitations and capabilities identified." The maximum effort was to be aimed at development and use of full Gambit potential. The NRO director told Greer, "...the name of the game is specific coverage of specific, known targets with stereo photography of the best possible quality." Mission criteria, he added, would be the number of priority targets photographed in stereo and the resolution of the photographs.[19]

Greer's confidence in the capability of Gambit was increasing, but he remained cautious. It was reasonable to assume that the camera was capable of being pointed accurately at ground targets, that orbit injection and orbit change maneuvers could allow coverage of different areas on the same flight, and that the resulting high resolution photography would be routinely recovered. The criteria settled on by the director of the NRO assumed these accomplishments and asked, if somewhat imprecisely, for considerably more: many photographs of many priority targets. To that time (1964), the selection of Gambit targets had been done manually; analysts decided which targets were to be photographed, how many frames were to be exposed over each, and the time sequence of the operation. The amount of film ordinarily available to be exposed was on the order of two thousand feet.

Two later developments in Gambit's operational capability were directly responsive to the requirement for the large numbers of aimed photographs. The first, which had long been planned for mission operations once R&D goals had been satisfied, was initially used on Gambit's fifth flight. It involved the use of computer techniques for target selection. The second was to increase the quantity of film carried, with a corresponding increase in the potential longevity of a given flight. Thirteen hundred feet of film had been loaded in the first Gambit mission. By the end of the program, almost 3400 feet were carried regularly.[20]

The first target selection technique used in the Gambit program, called TMPGP, was developed by Space Technology Laboratories (which later became the core of the Aerospace Corporation). It used data on priorities, resolution, and total number of areas of interest as inputs and then performed an exhaustive search of possible targets from target folder data in order to determine the optimal set of targets for a given mission. Because of the necessarily broad scope of the search technique, the program was bulky and expensive to operate, as well as being very time consuming. Program directors therefore sponsored a second generation program using dynamic programming techniques, which eventually replaced TMPGP. It was characterized by far more rapid and less costly operations but perhaps more important provided better satisfaction of optimal target selection criteria.[21]

Once the set of targets had been chosen, relevant data on them were entered into an event generation program which computed major orbital parameters as well as all commands necessary to carry out the selected photographic operations. Commands included such details as door openings and closings, crab angle, and film transport speed. Once the vehicle was on orbit it could be assumed that command changes would be required, either to correct for flight anomalies of various kinds or to incorporate late changes in mission objectives—as during a sudden crisis in the Middle East or Southeast Asia, for instance. These eventualities were taken care of by a set of command and control programs which allowed technicians at the satellite test center at Vandenberg to alter the sequence and nature of events on orbit. New target data were fed into the control computers; the output was in the form of commands to the space vehicle. In generating those commands, the program would also determine whether carrying out the new tasks would in any way degrade vehicle capability, either by requiring it to perform high risk operations or by overloading some subsystem to the point of incipient failure.

Such techniques were first employed on the flight of 25 February. The space vehicle separated from the Agena and entered the planned orbit, with an initial perigee of 96 miles. On the second revolution, telemetry indicated that roll and pitch gyros had not uncaged. It was faulty diagnosis, but not until somewhat later did it become apparent that only the signals were wrong, that in fact the flight control system was functioning correctly. In the meantime, flight controllers responded by sending a new "uncage gyros" command to the vehicle. Its effect was to cause loss of yaw reference and a steadily increasing yaw angle. (The yaw angle grew at the relatively large rate of 2.5 degrees per hour. Because the vehicle and the cameras were no longer at right angles to the orbital path, image motion compensation became steadily less effective. Resolution degraded from an initial 12 feet to an eventual 100 feet before the command and control system succeeded in negating and correcting the yaw angle anomaly, on revolution 18. Unfortunately, either communications between personnel at the Satellite Control Center Facility or the computer-controlled command system could not cope with the situation; in any case, a command to cut the film and load it into the take up spools had been transmitted and obeyed during the 16th revolution. Thereafter, Gambit number five performed some research operations which had been included as part of the primary mission, but photographic functions ceased.[22]

During early Gambit operations, interest in the feasibility of low level flights had become pronounced. Better resolution was the goal. But relatively little was known about the density of the atmosphere at altitudes of 70 miles. Gambit number five was photographically "dead," but something might be salvaged by having the vehicle descend to a lower altitude. Flight controllers quickly prepared and sent commands for three orbit adjustments. They demonstrated that the OCV could be successfully controlled and that operation at 70-mile altitudes was feasible. So notwithstanding the succession of command errors that led to a failure of mission photography, the mission was judged a success.

On the 34th revolution, the recovery vehicle was separated from the OCV and recovered. That event had originally been scheduled for the 51st revolution, but the extra time on orbit was to have been used for photography which had been precluded by the premature command to cut the film. In partial compensation, mission control extended the flight of the solo OCV from its original program of 32 revolutions to 49. During that period, a malfunction occurred in the BUSS (caused by the failure of a relay) which would have caused catastrophic failure in the event BUSS had been relied on for recovery of the RV. Other minor functional failures marred the flight, but for the first time instrumentation was sufficiently comprehensive to provide relatively detailed information on each of the anomalies. Detailed corrective modifications of the sixth Gambit followed analysis of the failure modes experienced during flight number five.[23]

The fifth flight had another distinction. It was the first for which an incentive fee arrangement had been in effect between the project office and General Electric, the contractor for the orbital control vehicle, which related fee to vehicle performance on orbit. Basically, GE would recover the costs of production plus an additional fee around a mission success midpoint. Penalties and rewards were on a pro rata basis between the two extremes. Because of various OCV failures in flight number five, GE had a full fee penalty assessed.

Incentives of that sort became increasingly important to Gambit as the program progressed. Although cost, performance, and schedule were all covered by the contract incentive clauses, cost was the principal early target. Gambit costs had substantially exceeded early program estimates, although later developments in satellite reconnaissance were to make that program seem quite inexpensive. In any case, the arrangement that took effect with the fifth Gambit mission provided that GE would retain as profit a large part of whatever underrun occurred, but would pay an equivalently large share of overruns out of fee. The fee variation associated with performance was only about half the size of the cost function variation. That difference was partly accounted for by the expectation that the orbital control vehicle would be extremely reliable, a notion strengthened by the first four flights. The schedule incentive was small. As might have been expected—indeed, as was intended, after the first couple of flights, GE concentrated effort on reducing costs, paying less heed to performance. Under the terms of the agreement, for instance, if GE delivered a minimum-cost OCV on schedule, and the vehicle was an utter failure on orbit, the contractor still would earn the full scheduled fee plus an incentive bonus. Of course that sort of arrangement would not be continued if failure became a major problem, but it was an interesting condition of the program and one that GE exploited—briefly.[24]

The failures of the fifth flight determined the operational assignment of the sixth. The mission would be the same: a three day flight to demonstrate full operational control and orbit adjustment capabilities, and continued investigation of low orbit operations. The primary mission would continue to be high-grade photography, but this was now a sufficiently hoary tradition to be accepted without notice in the formal statement of mission objectives.[25] The long awaited stellar index camera was originally scheduled for its maiden flight on the sixth vehicle, but qualification delays made it necessary to wait one more flight before that final important piece of equipment could be added to the Gambit package to make the vehicle fully consistent with original specifications.[26]

Even though the fifth Gambit had been less than successful from the standpoint of program personnel and the intelligence community alike, the system was making steady and remarkable progress toward full operational status. That circumstance was acknowledged by Greer's near-term plan for flights six through ten. The length of the flights was to gradually extend from three to five days, low altitude experimentation was to be ended, and optimal targeting procedures were to be developed so that the greatest possible number of high-priority targets would be photographed. Greer maintained that the plan was flexible, providing a "...deliberate approach to completing the development program. It can and will be adjusted to either unusual success or catastrophic failure." McMillan's concurrence in that general plan reached Greer, without additional comment, on 17 March 1964.[27]

The subtle increase of emphasis on obtaining operationally useful photography starting with flight six was in some respects a further acknowledgement of the pressure McMillan was experiencing from users of the satellite photography. The failure of Gambit five to return useful take was reflected without much subtlety, in measures that put Gambit six on its launch pad well ahead of normal delivery and checkout schedules. Only 15 days elapsed between the launch of the fifth Gambit and the readiness of the sixth. That striking accomplishment was made possible by energetic refurbishment of the launch station and by making detailed adjustments in delivery schedules for later Gambits and their subsystems. It was possible partly because the production, test, and delivery of satellite vehicle equipment had earlier been accelerated in order to provide a standby or backup vehicle as early as possible in the Gambit flight program. One effect of moving Gambit six forward in the schedule was to force a delay in the planned delivery of the reserve system; it had been slated for availability by May 1964, but the actions of March (in conjunction with an independent delay in the availability of the booster, the first SLV-3 "Standard Atlas" intended for use with Gambit) made June the earliest possible month for delivery.

Gambit six was launched shortly after noon on 11 March 1964. Orbit and separation were nominal. There were two major system failures during the flight, neither catastrophic. The BUSS failed for a second time due to overheating of a solenoid during the test cycle. (Test procedures were immediately changed.) The primary stabilization system operated nominally throughout the flight so that failure of the BUSS was of no mission significance.

The second major problem was malfunctioning of the roll jets, which caused orbital predictions to be inaccurate and led to in-track errors that made it impossible to acquire 25 percent of the planned targets. (The errors were apparently cumulative; the majority of targets not acquired were programmed for revolutions 38 to 41.) An additional problem was that some targets had been specified "by hand" rather than through the use of computerized event selection programs. While on orbit, it became clear that many of these were incompatible with the technical capability of Gambit hardware.[28] (That event brought on a short, sharp controversy over operational direction of intelligence gathering.)

Despite such problems, some 150 successful camera operations were commanded and took place of a programmed total of 229. After 51 revolutions, the film capsule was recovered and several orbital adjustments were made with the OCV in solo flight. With the first burn, the perigee was raised from 120 to 202 nautical miles, then returned to 95. In the next experiment it was reduced to 70 miles, then finally returned to 90. The vehicle had stayed on low orbit with a 70-mile perigee for a total of ten continuous revolutions. No temperature anomalies or other difficulties were registered, further demonstrating the feasibility of such a low orbit. The next step would be to operate the camera from such an altitude to determine if special problems would arise either from camera operation or in the quality of photography from low orbit.

Such events were of paramount importance to the Gambit program office, but the intelligence community they ranked considerably below the fact that Gambit six had returned substantial quantities of highly useful intelligence data. Earlier photography had generally been scant and, even when of good photographic quality, had provided information of slight operational interest. The photographs returned by the sixth Gambit permitted interpretation of additional details in already identified targets and confirmed the existence of targets in areas which had been classified as "probables" earlier. Good quality photographs were obtained on all but four of the assigned targets covered on the mission. Some of the photography was degraded as a result of snow or haze cover, but most was excellent.[29]

That achievement received far less recognition than it deserved, then or later, owing to continued concern for impending missions. But in fact, Gambit was only the second thoroughly satisfactory satellite reconnaissance system to reach operational status in a development effort that had been intense—a matter of extreme national urgency—for six of its ten years. Corona, the only predecessor system to provide much in the way of useful operational intelligence, had not recorded its first success until 15 launches had been attempted, only three of which could be accounted technically successful. All five of the E-series Samos payloads that had progressed as far as completed hardware had been cancelled, as had Lanyard, the repackaged E-5 camera system. The whole concept of readout that originally underlay the program had been dropped, and in the entire series of Samos-derived mission attempts that started in January 1961, only the E-1 and the Lanyard had returned photography in which photo interpreters could honestly express the slightest interest. As compared to Corona, E-1 had been thoroughly inferior, while Lanyard displayed various defects of system and product that made its cancellation inevitable once Gambit had demonstrated even minimal capability. In coverage, quality, and detail, Gambit photography obtained from the sixth mission represented a data acquisition success that could only have been matched, in earlier years, by aircraft operations so uncertain of success and so risky in a political sense that even in the worst stages of the several international crises of the early 1960's they were never seriously considered. And the only comparable successes of the next eight years were to be Gambit-3 and Hexagon. In that sense, the success of the sixth Gambit mission constituted one of six real achievement milestones in the first fifteen years of serious satellite reconnaissance development by the United States. The others were the original Corona, the Corona-Mural stereo system, the dual-capsule Corona-J, Gambit-3, and Hexagon. And two of those were improvements on existing systems. In that context, the first real operational success of Gambit in March 1964 could stand as one of the most remarkable achievements of U.S. technology in the first decade and a half of the space era.

With the success of the sixth flight, operational proving had been extended in several areas, helping to determine the character of the seventh flight. Two events were of particular significance. First, the stellar index camera was finally mounted on the satellite. Second, the low altitude tests seemed very promising for the future. Greer, King, and McMillan were agreed that they should take the final step of flying the recovery vehicle with camera operating at low altitude during the seventh flight. The operational plan called for one day of flight at a 90-mile perigee followed by two days at 70 miles. During the low flight, technicians at Vandenberg were to be prepared to adjust the orbit upward if any sign of unacceptable temperature rise appeared.[30]

The circulation of Gambit six photography through the intelligence community had one effect on Gambit seven plans that neither Greer nor McMillan appeared to have foreseen, and which they and the members of the Gambit project group justifiably regarded as both unwelcome and unwarranted. On 26 March 1964, two weeks before the scheduled launch of the seventh Gambit McMillan received from the Committee on Overhead Reconnaissance (COMOR) a request that data on the ephemeris and track of each Gambit mission be supplied to COMOR ten days in advance of launch so that, upon study, COMOR could add to the operational plan a set of photographic requests of its own.[31]

The request had its origin in two related circumstances, first that the Corona operations with which COMOR was familiar were target programmed "by hand," so to say, and second that COMOR was largely unfamiliar with the highly complex computerized techniques used to construct a Gambit mission profile. Underlying them, of course, was a COMOR charter that implied rights of target selection and recent history of disagreement about mission and target priorities— disagreement that stemmed in part from CIA beliefs that the Gambit program was insufficiently attentive to intelligence needs. In this instance COMOR was largely reflecting the CIA viewpoint.*

McMillan and Greer were in agreement about the importance of maximizing the amount of photography of priority targets on each mission. To that end, Greer had set in motion the technical effort required to plan each mission systematically, exploiting new computer methodology, so that the greatest number of highest priority targets would be photographed on each mission. That activity had only begun to bear fruit on the sixth Gambit flight. The idea of inserting even one randomly determined target in a mission so planned had nightmarish qualities. Such an insertion would almost certainly ensure that the mission was suboptimal—that some priority targets that might have been photographed would not be, and that those missed might, in sum, be more important than the single insertion that displaced them. Mission tinkering might not have that effect, of course, but it could.

Dr. McMillan turned over the problem of reply to General John Martin, head of the NRO Staff. Martin's response was swift and deft. He provided a primer on the technical capacity of Gambit and the means of utilizing it, explaining that orbital parameters were planned so as to be consistent with a target list which was a primary input for such computations and pointing out that the optimality of such a mission plan would be destroyed by insertion of last minute targets. He felt that COMOR should have learned about the technical aspects of Gambit operations from experience with earlier requests for special targets. He concluded that, "It is simply not possible to proceed on the basis of manual target determination as the mission progresses without substantial loss of potential intelligence take."[32]

_____
* J. Q. Reber, who later became Deputy Director, National Reconnaissance Office, was the chairman of COMOR and its acknowledged spokesman, he was also a CIA employee.

The problem went away. The later success of Gambit operations precluded its resurrection.

One point Martin did not make was that mission event planning was already having to cope with an excess of targeting requirements. The computer program then in use could absorb data on only 900 targets although some 2700 had been earlier specified as "eligible." This meant that some hand massaging would have to be done even before the optimization routine could begin.[33] And for Gambit number seven, there were other pre-launch problems. The launch date had already been slipped by about a week, but checkout was not going smoothly. On 16 April the mission was scrubbed because the checkout crew could not satisfactorily explain (and fix) command sequences anomalies which kept recurring in the test sequence.[34]

The seventh Gambit with stellar index camera aboard, was finally launched on 23 April. All mission assignments were successfully completed by the third day, after which the vehicle flew for an extra day, making Gambit seven the first four-day system. In particular, the two days spent in low orbit were uneventful; preparations for emergency reorbit maneuvers being unnecessary. The only major malfunction of the flight was registered in the failure of a component in the horizon sensor, although that relatively minor anomaly caused track errors of as much as four miles late in the flight. Some camera pointing problems resulted, with a consequent degradation of photographic quality after orbit number 42. Inspection of the photographs showed that the low altitude at which most had been taken enhanced their quality substantially. Stereo pairs were particularly good, registering a "best" resolution of two and a half feet. The stellar index camera had taken 663 frames, although many were of impaired value because of light flare in the stellar exposures and overexposure in the terrestrial. Initial estimates of the cost of fixing those deficiencies were too high, so as an interim measure, black felt material was affixed to the interior of the lens cone to suppress reflection.[35]

Growing recognition of inherent Gambit capabilities contributed to the next attempted perturbation of the planned program. The CIA expressed interest in operating the system over Cuba, although the continuing success of U-2 flights in that area would seem to have provided sufficient assurance that the missile crisis of 1962 could not recur. Attempts—and success—at shooting down U-2's would presumably signal the start of a new crisis. But Gambit (and Corona too, for that matter) was in its usual mission mode not well adapted to reconnaissance over Cuba, mostly because its flight plan was optimized for operations at higher latitudes, at different sun angles, and in another hemisphere. Night launches from Vandenberg could put the satellite over Cuba during daylight hours, as could launch from Cape Kennedy. Both options had severe drawbacks. A night launch from Vandenberg would create a mission capability limited to daylight operations over the Caribbean and some other areas where the U.S. had little or no intelligence interest. Daylight recovery near Hawaii would depend on successful orbit adjust and could not follow closely on photographic passes over Cuba. Facilities at Cape Kennedy were inadequate, and provision of minimum checkout capability would cost and take about nine months.

Basically, however, the Gambit camera-vehicle system was far less than optimal for the sort of Cuban coverage being considered. Gambit could, of course, take both mono and stero strip photography. But a mono strip photograph over Cuba would cover a swath ten miles wide and 600 miles long. Several such passes would be needed to cover the entire island and resolution could be expected to degrade because of small, cumulative in-track errors which would normally be corrected between targets. Using stereo pairs was not a better alternative: difficulties with slow settling times had still not been eradicated and computations showed that ten percent of time over target would be consumed by a single roll maneuver of 25 degrees. Finally, and ironically, the seventh Gambit flight had been a four-day mission, raising hopes of a five day mission shortly. Cuban coverage would need no more than a two-day mission, and that represented a costly disregard of the maximum technological capabilities of the satellite. Gambit obviously could do crisis reconnaissance if the need were sufficiently great, but it did not appear that Cuban reconnaissance qualified.[36]

Preparations for the eighth flight proceeded, unruffled by the distant flap over Cuba. Mission plans were as for the seventh flight except that the duration was to be four days. Low altitude flight would continue to be tried.

Launch occurred on 19 May just after noon, to be followed immediately by problems. After separation from the Agena, the satellite vehicle was injected into an orbit which was more than 30 miles below the planned 90-mile perigee. Initial perigee was 57 nautical miles—an altitude at which the satellite experienced 17 times the atmospheric density, for which it had been designed. In addition the vehicle was rolling very rapidly. As tracking equipment lost contact with the vehicle, most flight controllers felt the vehicle had no chance of surviving the disastrously low orbit, much less the rapid spin. But the time Gambit had passed over the first downrange tracking station, however, the spin had stabilized automatically. On the second revolution, normal procedures for orbit adjust maneuvers were successfully carried out, lifting the vehicle into its planned 90-mile orbit. For the next 13 revolutions the vehicle operated nominally, producing what would later be labeled "...high quality stereo photography considered by some to be the best imagery yet obtained from satellite photography." On the 15th revolution, however, the vehicle inexplicably lost all attitude reference. Just as mysteriously, it reappeared on the 25th revolution. Photography resumed, but in view of the various travails to which the satellite had been exposed, the mission was terminated on the second day instead of the fourth. Capsule recovery was uneventfully successful.[37]

Attempts to explain the attitude control failure of the eighth flight were dominated by concern about the initial low orbit. Fears that the high atmospheric density would destroy ablative materials and cause malfunctioning of various subsystems seemed to have been borne out when attitude reference disappeared. That failure was initially charged to the effects of atmospheric heating. But a similar failure occurred during the next flight, and later analysis of the flight track showed that it happened only when the satellite was over Antarctica. It became apparent that the horizon sensor (which maintained attitude reference by determining the position of the horizon beneath the vehicle and appropriately issuing roll command) could not distinguish between the temperature of Antarctic and the temperature of outer space—at least during winter in the southern hemisphere. Sensor redesign followed. Once the attitude sensor failure had been correctly credited to geography rather than atmospherics, there was little remaining doubt that the vehicle could withstand the rigors of extremely low altitude flight with no major deleterious effects. A bonus gained by the unscheduled experiment was a significant refinement of the standard model of the atmosphere, which it developed, was wrong for the 50 to 70 mile altitudes Gambit had penetrated.*

The ninth Gambit flight was little distinguished from the eighth except that it spent no time at 50-mile altitudes—planned or otherwise. It experienced the same problems of attitude control, however, and the effects were considerably worse. The best resolution obtained was on the order of fifty feet, making photography of little worth. Inadvertent exhaustion of orbital control gas owing to the attitude control problem was so acute that the BUSS had to be used

---

* Explaining where the correct data had originated was more of a problem than collecting and analyzing it; scientific satellites simply did not operate at 50-mile altitudes, aircraft could not go so high, and balloon data had been misleading. In the end, the "corrected" figures on atmospheric density were "surfaced" through NASA, with no real explanation of source, and apparently nobody noticed.

for capsule recovery. The best product of the flight was information about what had not caused the attitude reference problem of the eighth flight and where the correct solution should be sought. Three weeks after the mission ended, the project director, Colonel W. G. King was able to provide a definitive explanation for the entire episode.[38]

And, unhappily, starting with Gambit number nine there began a series of five missions which were generally poor in one or several ways. The best resolution obtained during the entire period from early July to the end of October 1964 was seven feet. Only 345 targets were photographed during a total period when five separate launches produced but five days on orbit. The tenth flight was to be, according to the schedule agreed upon between Greer and McMillan, the last of the Gambit development flights. The generally good records of the sixth, seventh and eighth Gambit missions had roused sanguine expectations of returns thereafter, yet the three men most responsible for Gambit, McMillan, Greer and King, were too canny about research and development to count overmuch on a run of good luck.

The hardware for the tenth Gambit flight had been modified in response to earlier problems. In particular, a new backup electromechanical device had been installed to operate the primary camera door in case the pneumatic system failed. Gambit number ten was also the first to use the new Atlas Standard Launch Vehicle (SLV-3). But early in the flight, an electrical failure, on 14 August, followed by a blown fuse, induced failure of the stellar index camera, making exact location of primary photographs difficult. From the 19th revolution the command programmer could not be loaded, the result of either a parts failure or poor contact in a coaxial connector for the harness between decoder and programmer. After recovery and photoanalysis, poor resolution suggested a misalignment of the photo slit. It was only later that engineers discovered that the camera was out of focus because of a malfunctioning temperature sensor which forwarded incorrect temperature compensation data to focusing devices. (The problem was not identified until it recurred on the eleventh flight.)

That eleventh flight, launched on 23 September, was marred by a host of problems. The focus error, incorrectly diagnosed from the previous flight recurred. In addition, two separate instances of valve contamination were identified: one valve was inhibited from opening, and the other remained open, allowing a slow leak. A new problem occurred with the stellar index camera. Camera access doors would open only partly because of weak springs (improper heat treating) and incorrect door clearances. The saving grace of the eleventh flight was that pointing accuracy of the camera proved to be superior to all previous experience with Gambit.[39]

While the ninth, tenth and eleventh flights returned only poor and small amounts of photography, the next two missions returned none at all. The twelfth flight, an 8 October 1964 launch, experienced an Agena failure and was the only Gambit flight in the original series that failed even to orbit.

Two weeks later, the thirteenth Gambit had been hurried to readiness. The intelligence community had gone almost four months without adequate coverage of important targets and expressed understandable uneasiness. Program managers were reasonably confident that there was sufficient information about hardware shortcomings to support adequate corrective measures, however. On 23 October the thirteenth bird was launched and successfully injected into orbit. The four-day flight went well, all subsystems working perfectly as far as the ground crews could ascertain. On the 67th revolution, the command to retrofire was sent to the recovery vehicle. The reentry vehicle separated from the orbital control vehicle but thereafter—nothing. The backup systems available on the OCV could not be used. Natural orbital decay finally brought the capsule down on its 93rd revolution, but reentry point could not be accurately calculated and it was lost.[40]

Several consequences arose directly in that sequence of failures. Dr. McMillan reiterated, in their aftermath, that the principal program objective was to achieve one successful mission every 40 days. In line with this objective, the Secretary of Defense approved the procurement of an additional four Gambits during fiscal year 1966 (raising the total on order to 16) and an additional three for the next year, raising that total to fifteen. In addition, McMillan ordered the curtailment of all supplementary development of Gambit in an effort to improve the chances of success for the next few flights.[41]

A problem which was to recur the following year was the low quality of workmanship on the orbital control vehicle. The project office concluded that "... poor discipline in factory and field... by the SV contractor..." caused black-box failures at an unacceptably high rate. General Electric responded to such prodding by agreeing to a series of remedial actions: a reduction in overtime worked, slippage of delivery schedule to allow system modifications to be completed at the factory rather than in the field, and a general tightening up of personnel control and training.

The terms on which agreement was reached indicate that GE was very concerned about the incentive scoring system. The contractor felt, for instance, that the introduction of extra payloads decreased his control over the chances for success—and a high incentive score. After a string of almost perfect scores, the ninth and tenth flights cost GE penalties. Oddly enough, despite several findings of faulty workmanship, the eleventh flight brought nearly the maximum fee to the contractor. There were clear indications that the contract incentive structure required overhaul.

General Electric made serious efforts to improve production, control and testing. Flight results suggested, for instance, that thermal vacuum testing had been inadequate; both its intensity and its duration were enlarged. Vibration testing was also changed. Instead of "before and after" tests, "operation during vibration testing" was required. Faults that earlier had gone undetected were thereby identified.[42]

During the early months of 1965 the horizon sensor problem was finally solved. The cause of loss of attitude reference had already been identified as the inability of the sensor to distinguish between earth and space during winter months over the South Pole. The first response to this difficulty was to attempt the development of a more sensitive device, but initial estimates of the cost and time required proved low and the real probable cost unduly high. Spurred on by this, by study on the terrestrial mechanics of Gambit flights, and by the approach of warmer weather at the south pole, program managers found a cheap solution. No targets of any value existed over the south polar regions, so the easiest answer was to let the vehicle coast over the area un-stabilized. Once it returned to warmer latitudes, the horizon sensor could be reactivated and attitude control regained. The solution saved more than the money required to develop a better sensor; it also permitted a significant saving of orbital control gas.[43]

The fourteenth flight of Gambit began a period of successful operation which was to continue through the summer of 1965. During the period, there was a significant increase in the amount of photography produced and the number of targets photographed, as well as a steady improvement in resolution. The fourteenth flight was the least successful of the lot, being aborted after only one day on orbit, and clouds interfered with photography during much of that day, but the recovered film registered a best resolution of 2.1 feet. The BUSS control system was altered: Gambit number 14 incorporated several new features. BUSS commands were changed from single to double tone, and an address command was inserted in the BUSS programs. The first change resulted from the fact that fishing vessels frequently used the BUSS frequency for communications and occasionally transmitted the critical tones, triggering spurious commands of the BUSS. The second change was introduced in anticipation of having more than one vehicle on orbit simultaneously.

Gambit number 14, launched on 4 December 1964, operated successfully through the first eight revolutions. During the ninth, however, battery overheating was followed by a loss of stability. The vehicle was recovered via BUSS during the 18th revolution, providing a successful test of the new command coding.[44]

A month and a half later, the fifteenth Gambit was launched. It was the first vehicle to incorporate the subsystems generated by GE's changed quality control process. Launched on 23 January 1965, Gambit 15 went through separation and injection and the first few revolutions without untoward incident. But thereafter, three main heaters malfunctioned and temperatures in critical sections of the space vehicle degraded throughout the flight. The result was that although photographs taken early in the flight had resolution of as good as two feet, they gradually diminished to ten feet by the fourth day. The vehicle was recovered during the fourth day, completing its planned mission successfully.

The previous three flights had each been scheduled for five days of operations and none had lasted longer than one day on orbit. By the time the fifteenth flight was launched, the seriousness of the problem was such that the longevity aim was actually reduced to four days.[45]

Despite heater difficulties and a serious failure of the stereo mirror, Gambit 15 could be considered successful. The photographic take, of variable resolution, covered 688 targets—more than any previous flight and more than the five preceding flights together.

The sixteenth flight began on 12 March under the same operational plan as the fifteenth. Again, a high volume of intelligence material was produced, but as there was no heater problem the photographs were of uniformly high resolution. The major drawback was that on revolution 16 the stereo mirror stuck again, allowing only mono photography.

The stereo mirror problem was finally pinned down by Eastman Kodak engineers as a result of data retrieved from Gambit 16. The fault was improper relay sequencing to the stereo servo which could cause arcing, the consequent welding of relay contacts, and freezing of the mirror in whatever happened to be its position when the weld occurred. Correction was relatively simple once the cause had been identified: the command sequence was changed. Although other relatively minor problems of command transmission and decoding affected some of the photography, the flight returned a high volume of useable intelligence and was accounted successful.

The seventeenth flight of Gambit was something of a watershed for the program. It incorporated the products of all the hardware and procedural changes of the past year, a set that extended from improved testing and production control techniques to the reduction of flown on the vehicle. Two successive four day mission successes had increased confidence that the five-day mission objective was now achievable, and correction of the sticking stereo mirror problem eliminated the last known major technical defect of the system.

The mission began on 28 April. Except for one malfunction, it was superb. Photographic coverage increased to 992 targets, best resolution reached two feet (equaling the earlier "best"), and operation of the stereo mirror was uneventful, allowing 180 stereo pairs to be produced in addition to a high volume of mono photography. The single malfunction was in the primary camera door actuator, apparently a result of binding between the door and the opening or some nearby harnessing. The backup system overcame the difficulty, however, and the outcome of the mission was not affected.[46]

The 17th Gambit was distinctive in another way, apart from its superlative flight performance and record intelligence return. It incorporated extensive new failure mode detection and diagnostic devices and associated telemetry. In some respects it was odd that such comprehensive instrumentation first appeared on the 17th Gambit vehicle. The greater need would appear to have passed, particularly if account were taken of the exceptionally good performance of the 17th vehicle and its two immediate predecessors. But as had been the case with Corona, the only earlier satellite reconnaissance system to provide useful intelligence returns, Gambit had no sooner demonstrated that it could satisfy (and in some respects, exceed) the original program requirements than proposals for modifying it to produce still better intelligence began to surface. The feasibility of six- to eight-day missions was being seriously evaluated—and Gambit-3 was midway between first contract and first flight.* Those specific developments and several subsidiary aspects of Gambit evolution lent both respectability and urgency to the effort to obtain more definitive information on potential reliability enhancement, and that in the end was the object of the instrumentation effort. But it was not merely capability enlargement that encouraged attention to flight instrumentation; project officers were painfully aware of the possibility that the Gambit system could experience another plague of minor and major malfunctions that would inhibit its immediate usefulness, and experience of the recent past had clearly demonstrated that incorrectly diagnosed malfunctions tended to recur. (The attitude stabilization problem and the camera door difficulties were two painful reminders of the need for adequate instrumentation and diagnostic capability.) As it happened, the incorporation of that additional instrumentation was providential; Gambit was indeed about to experience another set of flight difficulties not unlike those of flights nine through thirteen and for reasons very much like those behind the earlier difficulties.[47]

The 18th Gambit was launched on 27 May 1965 with the assignment of performing a five-day operational mission. Apart from two non-significant functional problems (one being a recurrence of the earlier door-actuator failure), it experienced no on-orbit difficulties. The quantity of returned film was again larger than on any earlier flight, and resolution again reached a "best recorded" level of two feet. Two successive operations so superlative made it appear that Gambit had indeed matured, that it was a fully reliable operational vehicle subject only to the random minor disabilities inevitable in so complex a system.

As though reasserting its rights to perversity, Gambit number 19 was the complete antithesis of its immediate predecessors. Launched on 25 June 1965, it experienced a massive short circuit during ascent. The electrical failure completely disabled the stabilization system and the flight programmer. Either event was catastrophic. Notwithstanding the strenuous efforts of flight controllers to regain command, the vehicle remained unstable and uncontrollable during the 18 revolutions it logged. It was recovered via the Back-Up Stabilization System on the 18th revolution. Usable output from the flight was nil.

---

* As earlier explained, in order to distinguish between the two programs the designator "Gambit-3" will hereafter be used in this manuscript to identify the two-capsule, long-lens system that, after June 1967, was formally known only as Gambit. A similar distinction has been made elsewhere: the terms Corona-Mural (or Corona-M) and Corona-J have been used here even though contemporary documents did not distinguish among the several variants of that system. It is perhaps worth noting that even after the original system disappeared, discussions among satellite program participants generally included references to "G-cube" rather than "Gambit," or "G."

The initial diagnosis of the failure was that it had been induced by the extremely high vibration associated with the boost phase of the mission. The short circuit, which occurred on the ground side of the direct current power supply, opened the power supply input filter, disabling the stabilization subsystem.[48] More disturbing, the ultimate cause of the failure had to be either contamination or a faulty part—which immediately suggested that quality control and testing procedures at General Electric had been grossly inadequate.

On that unhappy note, Major General Robert E. Greer left the program, and the Air Force, on 30 June 1965. He, more than any other individual in or out of government, had been responsible for instigating the Gambit program and for carrying it to its mid-1965 level of proficiency. With Colonel W. G. King, he had been personally responsible for all of the major, and quite difficult, technical and management decisions that marked the program's first five years. (Gambit had been invented, in a sense, in the summer of 1960, although it had not taken form as a system program until December of that year.) By the only valid standard of comparison then available, the early Corona program, Gambit represented the most comprehensive and striking success yet achieved by the American reconnaissance satellite program. It had recorded its successes earlier and with greater regularity than the early Corona, and the returned photography was in its own special way of equivalent or greater value. And by the summer of 1965, there was abundant evidence that Gambit could be improved at least as markedly as Corona had earlier been improved, with a consequent equivalent benefit to the overall satellite reconnaissance program. On balance, the achievement was quite remarkable.

Brigadier General John L. Martin, Jr., who had earlier headed the NRO staff in Washington but who had most recently been Greer's deputy, succeeded to Greer's post in Los Angeles. Virtually from the day of his accession, he was confronted with the question of whether to proceed with the next scheduled Gambit launch on 9 July or to delay the mission in order to revalidate the probability of mission success. Given that the failure of Gambit number 19 could well have been caused by a random breakdown of the quality control, inspection, or testing procedures, Martin decided to proceed in accordance with existing plans. In any case, he was immediately confronted by a massive problem in contracting and procurement.

Nearly two months earlier, program officers had advised Lockheed Missiles and Space Company of their increasing distaste for the high prices reflected in Lockheed bids on new Agena vehicles. Costs were much higher than for earlier deliveries of approximately the same equipment. Procurement officers concluded that Lockheed was negotiating to protect a position rather than "in good faith." They were also concerned that Lockheed might be maintaining a large reserve pool of engineers who did not work on Gambit but were funded by that contract. Even more than was usually the case for a sole-source supplier to the government, Lockheed was in a very favorable situation for negotiating follow-on procurement. Agena production had continued at a regular rate for years and bid fair to continue for several more. NRO people had long since explored and discarded as unfeasible the possibility of establishing an alternative production source. It promised to be an extremely costly course, and one involving considerable technical risk. Nor, in general, could Lockheed be faulted for inferior Agena performance. Although some quality control problems had occasionally appeared, the Agena was widely regarded, at the time, as a reliable vehicle—something of a contrast to the more troublesome GE orbital vehicle, which was the object of considerably more immediate concern on that score.[49]

The twentieth flight of Gambit slipped three days, to 12 July. The launch was a prompt and total catastrophe. The Atlas flight programmer shut down the sustainer engine prematurely and the Agena and its payload followed a ballistic trajectory to impact in the Pacific some 680 miles south of Vandenberg.

Although the Gambit-20 mission had been a complete failure, the fault was almost unique. Atlas boosters rarely malfunctioned so thoroughly. The OCV and the Agena, more characteristic sources of program difficulty, had not been given a chance to demonstrate their capability. Nevertheless, the entire Gambit system was subjected to new and more stringent test and inspection procedures starting with mission twenty-one. In particular, OCV components were subjected to x-ray inspection, and second, the intensity of vibration testing was increased by 30 percent. Both of these measures led to the discovery of faulty OCV components, and in six other instances they were attributable to insufficient quality control and inspection during manufacture and assembly by GE. Power supply subassemblies, which had caused the June failure, were re-examined in detail, with the result that several instances of the incorrect application of thermal grease were detected.

The correction of such defects and the redefinition of mission objectives caused a schedule slippage of one week. On 3 August 1965, the 21st Gambit was launched. It achieved orbit without difficulty but the AC/DC power converter in the OCV promptly failed, resulting in an immediate and permanent loss of stability. No acceptable photographs were recovered. Gambit number 21 thus became the third in succession to experience catastrophic failure.[50]

Understandably, the intelligence community was becoming increasingly concerned about the gap in detailed coverage of the Soviet missile program that three successive Gambit mission failures had caused. The last good high-resolution photography of what was known to be an intensive Soviet ICBM buildup had been recovered in May; owing to the rapid depletion of Gambit hardware, launch schedules could not be accelerated and not until at least late August would it be feasible to attempt another mission. (In the event, a September launch date proved to be the earliest that was achievable.) The minimum program goal of one successful Gambit flight each 40 days had gone by default. Nor was Gambit the sole—or even the paramount—concern of the NRO during the summer of 1965. The Washington staff had been involved in institutional bickering between the Pentagon and the CIA which in September 1965 led to the departure of Brockway McMillan, for more than two years the Director of the NRO. Although the possibility that the NRO might be entirely abandoned as an instrument of national reconnaissance policy was dispelled by the appointment of a successor to McMillan (Dr. Alexander Flax) and by the issuance of a new NRO charter, the whole of the reconnaissance program was in some disorder. Corona operations had been reasonably successful during that summer, only one major mission failure having occurred in three flights, but Corona did not return the detail that intelligence analysts had begun to expect and interpretation of Soviet force status had become heavily dependent on information elicited from Gambit photography. Some part of the institutional infighting of 1965 was occasioned by disagreement over the management of the Corona program and some of the Corona project people on the West Coast were convinced that a serious failure of Corona operations could result if the authority for technical and operational control of that bifurcated activity was not promptly sorted out. Although in retrospect that appeared to be no more than a minor possibility, it contributed to uneasiness on both coasts. And finally, an extended controversy about the nature and timing of a replacement system for Corona, and perhaps also for Gambit, was complicating plans for the continuation and improvement of both systems.

Flax had to turn his attention to several of these issues almost simultaneously; his immediate reaction to the Gambit problem was to suggest study of the possibility that "twenty to thirty" Gambit launches might be conducted each year. (At that time, increasing the schedule from 13 to 15 annual launches was occupying the project office; a 20-launch-per-year program would require about an 80 percent expansion of production capacity—a considerable undertaking.) He was also very attentive to measures initiated by General Martin that were intended to improve markedly the general quality and reliability of delivered Gambit subsystems, particularly the Agena and the orbital control vehicle (which had, on balance, provided most of the program difficulties, the Atlas and the camera system being infrequent offenders.)

General Martin's response to his problem of the moment—Gambit was to obtain approval of his proposal that the next scheduled Gambit launch be delayed by a month to permit more comprehensive testing and the incorporation of whatever correctives might be needed to insure mission success. He, too, was keenly aware of widespread uneasiness about Gambit's potential. The decision to delay launch was not lightly taken; it guaranteed, at the least, that the next delivery of Gambit photography would occur at least a month later than had earlier been expected.[51]

The more extensive thermal vacuum and vibration tests being conducted by GE were uncovering large numbers of faulty parts and assemblies, frequent contamination, and other defects of workmanship. But to Martin's relief, nothing seemed to be inherently wrong with the design of the OCV or its interfaces with the Agena and the camera. Consistent with these findings, thermal vacuum testing was extended to 48 hours and expanded to include the entire OCV. Similar tests were also applied to 24 critical components. Vibration testing was also expanded to include complete systems with equipment operating during the period of vibration. Inspection teams began tear-down of 25 mission critical components on each vehicle, searching for contamination and bad workmanship—and finding more than enough to justify the time they spent. The vibration tests were enhanced by improved monitoring devices so that part failures or malfunctions could be more easily identified. Components which had earlier been extensively modified were subjected to complete requalification. (Many components little resembled their original, qualified format.) All inspection procedures were sharpened. Finally, because some modules reworked following identification of a failure became even more prone to failure, GE began an effort to decrease frequency of reworking.

While some of these changes were routine enough and cheap enough to be continued thereafter, others were extraordinary measures adopted temporarily in response to what was widely regarded as a transitory crisis. The Air Force lacked the resources to support such a complex process of test and checkout through the life of an operational program. Recognizing that circumstance, General Martin began to plan for the adoption of a novel contract incentive scheme he had originated earlier, while serving as Greer's deputy. It was pointed more at GE than Lockheed, at first, because the failure of the 21st Gambit had resulted from the third catastrophic OCV failure in five flights. Martin's point of attack was the incentive fee contract with GE. His study of the existent contract incentive provisions led him to conclude that they were most appropriate for the development stages of the program and decidedly inadequate for the operational phase (which had presumably begun with the tenth flight).

The incentive structure earlier installed emphasized the importance of cost over operational performance. It had been, at least in part, prompted by lost control problems characteristic of the early Gambit program. But it also reflected the experiences of Greer and King with previous satellite reconnaissance programs in the older Samos series. With few exceptions, they had incurred major cost growth. King's reputation for bringing high-cost, high risk programs under control was highly regarded, and in assuming control of Gambit, he had done precisely that.

But Gambit was no longer a development-focused activity, despite the continuation of engineering improvement activities. To the extent that the nature of satellite reconnaissance vehicles would permit, Gambit was a production item—withal one that little resembled the usual military article.

A second point seems to have been the expectation that as the Gambit program continued, the contractors, as a matter of course, would strive to earn the bulk of the performance incentive fee. The original contract incentive program perfectly reflected such considerations and beliefs.

The incentive structure had three major parts: schedule incentives, cost incentives and performance incentives. The schedule consideration was in fact a disincentive for late delivery of the vehicle. The maximum penalty for late delivery was set as a per day amount, with a ceiling. On the other hand, a cost overrun carried fee penalties. Since the return to capital is computed by dividing fee by gross cost, that arrangement meant that the rate of return on gross costs was a variable function of vehicle cost, dropping sharply for overruns and rising sharply for underruns.

The performance incentive, unlike the cost incentive, was linear, being unrelated to the gross outlay for a given vehicle. A scoring system was devised on a scale from zero to 100. The critical region initially fell between 65 and 95, but these numbers increased as the system became more fully operational. A score of 80 was the breakeven point where no incentive fees were either gained or lost. For scores above or below 80, the fee changed in proportion to the change in the score. The maximum gain or loss in fee that was possible under such a system was on the order of half the amount that could be gained or lost via the cost incentive. To any rational contractor, that arrangement was an imperative to worry about cost far more than about performance.

One result of the bias was that GE was motivated to delete as many control and test procedures as possible in order to save money in the production of the vehicle. If, for instance, the deletion of a given test procedure had the same effect on reducing cost as on decreasing the probability of a failure, it would rationally be deleted, since half of the savings would be returned as an incentive fee on cost—over and above any penalty for inferior performance. Because that accommodation also reduced the capital outlay of the contractor, the resulting fee increase would be proportionately larger than the fee differences arising from flight performance bonuses or penalties.

Taken to its logical extreme, the formula could result in the delivery of a minimum-cost vehicle which failed catastrophically, but nevertheless earned a premium over and above the standard fee. The rate of return on invested capital in that case would be greater than about twice the normally acceptable return on fairly risky investments by private firms.

General Martin's arrangement left the schedule incentive essentially unchanged, but radically altered the relationship between cost and performance incentives. The new system paid no bonus for a cost underrun, a reflection of the belief that the cost of a vehicle built at that relatively late stage in the program could be estimated rather precisely. The maximum penalty that could be incurred for cost overruns was about what it had been. The major change was in the performance incentive. It no longer made sense to sacrifice performance for cost savings because costs below negotiated price brought no incentive fee, while performance shortfalls would reduce the fee at a much more rapid rate than before. Furthermore, even with an overrun of more than 25 percent, perfect performance meant a fee bonus. Most military procurements of the period were suffering from overruns at least as large as 25 percent, so no rational contractor would quarrel with the conjunction of a large price increase coupled with an incentive fee.

In retrospect, General Martin's incentive system represented probably the most significant non-technical accomplishment of the Gambit program. It recognized the fact that contractor performance could, in some instances, be "fine tuned" to the objectives of the contracting agency. In this case, shifting the focus of the incentive system from development to operations had precisely its intended effect—to judge Gambit missions to which it applied (number 24 and after).

Hindsight illuminates what General Martin saw: the contract performance of GE during 1965 steadily deteriorated, while fees did not. It seems clear that GE was reacting to an inappropriate incentive structure. Perhaps the change could have been made earlier. But the signs that seemed to stand out clearly after the fact—workmanship deterioration, faulty inspection, inadequate testing, and catastrophic failures resulting from such causes rather than from basic engineering design problems—were not readily detectable in the normal events of the early Gambit program. The success of early Gambit flights did nothing to make the identification of these problems any easier; when all goes reasonably well, prophets of doom have small voices.[52]

The introduction of expanded checkout procedures affected the program immediately, influencing the flights of Gambit 22 and 23. But those two vehicles were not subject to the provisions of the new incentive system. Nevertheless, changes in vehicle testing and quality control were rewarded by a successful flight for Gambit 22. That satellite was launched on 30 September 1965 after a lacuna of almost two months. The flight plan specified five days on orbit, but the excessive use of stabilization gas by the fourth day lessened vehicle stability so much that the capsule was called down on the 67th revolution. Some heating problems associated with the direct current power supply caused flight planners to reduce the battery charge below normal, but overall, the flight was a striking success. Some were photographed and for the first time the two- to three-foot resolution required by the specifications for the vehicle. This photography was the first intelligence material produced by Gambit since the end of May 1965.[53]

The Gambit vehicle for flight number 23 differed in some important respects from earlier versions. Most important were the changes that made it the first six-day vehicle. They included the installation of a sixth battery and a 12 percent increase in Freon—control gas—loading. The six-day capability had basically been made possible by the research on and development of thin-base film, permitting a 20 percent greater quantity of film (by area) to be carried without any increase in the size of loading or take-up cassettes. Providing stabilization control for six rather than five days required several minor changes (computer programming for instance) but increasing control gas capacity was the most substantial.

Owing to several circumstances (the new test and inspection regimen, the modifications needed to support a six-day mission, and delays in GE delivery schedules caused by correction of defects uncovered during checkout), the OCV for Gambit mission 23 reached Vandenberg seven weeks later than planned, arriving on 14 October. Nonetheless, it was launched on schedule, on 8 November. The launch was called "a good job" and injection went as well. But the excessive use of stabilization gas which had been experienced on the previous flight recurred; all gas was exhausted by the tenth revolution. The cause was leakage from the high-pressure regulator. The failure mode was peculiar in that it also caused thrust control valving to lose effectiveness, so that stabilization control vanished earlier than would have been the case had the leak occurred elsewhere. Photographs were taken during the 18 revolutions of the flight, of which ten were stabilized. Resolution was so poor that it could not be measured.

The gas pressure failure on Gambit 23 led to a detailed design change which became effective immediately before Gambit-3 replaced the original version, but it also lent further impetus to the continuing shift from pneumatic to electro-mechanical devices for Gambit-1. Owing to the excellent instrumentation carried on the 23rd vehicle, considerable data applicable to engineering improvements for application to later Gambits were obtained.[54]

Gambit number 23 was a turning point of another—and welcome—sort; it was the last of the lot of generally imperfect vehicles flown in 1965. The next ten flights were to be almost unqualified successes. The combination of quality control enhancement and the new contract incentive system became operative for the 24th and later Gambits. Each of the next ten flights experienced malfunctions of one sort or another, but none could be called a failure in consequence of those malfunctions. Those missions extended from January to October 1966, Gambits being launched about once a month during that period. They routinely returned photographic intelligence of high quality, covering more than a thousand targets on each flight. One flight returned photographs with a "Best Resolution" ranged from 2-5 feet for the entire series of missions.

Major system anomalies in that set of missions included two malfunctions of the stereo mirror, two stellar index camera breakdowns, and four instances of orbital control difficulties. None was grave enough to imperil mission success, although each had the effect of degrading total mission success in some degree. The OCV experienced no malfunctions of any sort during the first five of the ten flights, but four of the last five were so marred. The new testing procedures were extremely effective, locating and identifying several potentially catastrophic failure modes well in advance of launch. If only half of the major problems so identified by the new procedure had gone undetected by launch time, the majority of the missions would have failed or returned much degraded photography.

Beyond these matters, Gambit registered several notable achievements during the period. Although the first two missions had six-day capability, they were limited to five days because of minor system malfunctions. The third flight with six-day capability, however, flew for something more than six days, registering 99 revolutions and the remainder of the planned six-day flights performed as scheduled.

On the third anniversary of the flight program, 12 July 1966, the first vehicle with eight-day capability was launched. It was recovered on the 20th, having precisely satisfied the extended mission requirement. A comparison of the two flights, three years apart, had some very interesting elements. The first Gambit had been quite successful, in its own way, even though limited to what was predominately a research and development function. But in the interim, Gambit had extended its longevity (one to eight days on orbit), had increased the number of targets (from three to 1636), and had improved resolution (from 3.5 feet to 2.5 feet). And except for longevity, the initial eight-day Gambit was not particularly distinguished; it had the poorest resolution of the ten-mission set in 1966 and returned 20 percent fewer target photography than the best of the lot.

The first eight-day mission was the 30th Gambit. The 31st had originally been scheduled for launch eighteen days later, on 30 July 1966, but the first wildly successful Gambit-3 flight intervened. Gambit-3 had more than twice the film capacity of its predecessor, two recovery capsules, and a lens with a focal length of 160 rather than 77 inches. Gambit-3 was so extraordinarily successful in its initial operation that the need for launching the 30th in the original Gambit series on its original schedule completely vanished. Indeed, the returns from the first Gambit-3 mission so overloaded the film processing and interpretation capabilities of the National Reconnaissance Program that had the 31st Gambit operated with its usual effectiveness there would have been no timely way to exploit the intelligence return. Whatever else Gambit might have achieved in its first three years, it had completely suffocated—in film—early complaints from the intelligence community that not enough data were being returned. Now the problem was an excess of photography, a surfeit that would continue for a considerable period.

The string of successes that had begun with Gambit-1 number 24 was broken at ten. The cause was almost as infuriating as the flight was disappointing. The manufacturer of the explosive charges that removed the camera doors had changed pyrotechnic specifications without properly advising those responsible for altering the remainder of the camera door actuation sequence. In consequence, the door itself was not altered to accommodate the new changes and could not be removed once the vehicle was on orbit. Only stellar index photographs could be made. A new manufacturer was found, the door was altered appropriately, and provisions were made for door removal even if only one of the charges operated correctly. No further problems of that sort arose.[55]

Gambit number 34 was the last instance of catastrophic failure in the original Gambit-1 flight program, although less serious malfunctions continued to occur. Gambit number 35 suffered from an excessive roll time at low rate, causing degradation both in the number of targets that could be photographed and in the resolution (2.5 feet). A programming error caused selection of the wrong shutter slit during revolutions 7 to 25 of the 36th Gambit mission, with approximately the same effect. The last two flights, Gambits number 37 and 38, registered no problems of any consequence, an outcome that had marked only one other operation in the entire program (number 31). Those flights were also distinguished for photographing the greatest number of targets on mission 38) and for the best resolution obtained in the original Gambit-1 program.[56]

The last three flights extended from February to June 1967. By the time the last capsule was recovered on 12 June 1967, most of the remaining resources of the Gambit program had been dispersed. The distributed residual included four orbital vehicles and four Atlas launch vehicles, plus several cameras and Agenas.[57]

The original Gambit-1 program had been completely phased out by the end of June 1967, not quite three years after the first of its 38 missions. The name survived: "Gambit-cubed" dropped its suffix and thereafter was known as "Gambit." The newer system had by that time completed eight operations, mostly successful, and was returning film images with resolutions that were the best ever obtained from the 77-inch cameras used in the older reconnaissance satellite. Had that not been the case, had Gambit-3 been troubled by major problems of operation or resolution, four additional missions using the older system could have been mounted. They proved unnecessary. In a special sense, the older Gambit was the victim of its own success. The best of the original system had been carried over to the new, and progress in technology combined with simple economics to make continuation of the older system an exercise in inefficiency. Gambit-3 cost somewhat more, but one

Gambit-3 mission returned more than twice as much film, at better resolutions, and of more targets than the original could ever manage.

Even after Gambit-1 became one of the several casualties of technological obsolescence in the American satellite reconnaissance program, the system added another "first" to its considerable record of accomplishment. It became the first of the satellite reconnaissance systems—the first of all clandestine reconnaissance systems—to leave behind both a careful historical record* and a full set of system "hardware" deliberately stored against the day when it could be openly displayed.

---

* Somewhat sketchy historical accounts of the early Samos program appeared in the Air Force histories prepared at Wright Air Development Center (later the Aeronautical Systems Division of the Air Force Systems Command) in the mid-1950's but even then access to program details was difficult to acquire. Still sketchier records appeared in early chronological summaries of activity at the Ballistic Missiles Division (later the Space and Missiles Systems Organization of the Air Force Systems Command) from about 1956 until early 1960. The first serious attempt to write a history of any such program was sponsored by Major General (then Brigadier General) Robert E. Greer in 1962. He arranged to have Robert Perry, at that time the Air Force historian for the Air Force Space Systems Division, assigned to his organization, the Special Projects Office, on an informal, part-time basis. Greer's expressed purpose was to insure that accounts of the increasingly complex Air Force reconnaissance satellite program were prepared before the vital records disappeared. His support was continued and enlarged by his successors (Generals J. L. Martin, W. G. King, Lew Allen, and D. D. Bradburn). The activity to be covered by the history also expanded substantially, largely at the urging of Colonel Paul E. Worthman, an early Corona program manager and subsequently the long-term chief of plans for successive heads of the National Reconnaissance Office staff in the Pentagon. Perry continued to work toward a comprehensive satellite reconnaissance program history after leaving his Air Force position to join the research staff of the Rand Corporation in 1964, and became a contract historian after transferring from Rand to Technology Service Corporation in 1972. He was briefly assisted by W. D. Putnam, another former Air Force historian employed by Rand, in 1969-70. Bureaucratic considerations (the "blue suit" Air Force would not agree to the expenditure of Project Rand contract funds or such work) interrupted the preparation of the history between 1969 and 1973, and relatively little was done in the years 1967-69 because of Perry's primary commitment to the Rand Corporation assignments. The work was taken up again late in 1972 under contract between the Special Projects Office and Technology Service Corporation, at which time Robert A. Butler, a consultant with that firm, became a collaborator. The product of that spasmodic work over a period of ten years (to the time of this note) is this manuscript—which includes coverage of the background of Samos, the several E-series Samos programs, Corona and its descendants, Gambit, the evolution of the National Reconnaissance Office and its early activities, and related issues and programs To the best knowledge of the present authors and present and past members of the NRO staff, there is no formal history of any other reconnaissance program ever conducted by the United States. A CIA sponsored history of Corona was nominally in preparation late in 1972, and apparently some effort within CIA has been devoted to preserving records of the Idealist (U-2) and Oxcart (A-12) aircraft programs, but that represents the sum of such history. The ancestor of all such programs, the balloon-carried reconnaissance camera system of the mid-1950's, appears to have disappeared from the records. Given the volume of documentation of reconnaissance program activity by 1970, that is unlikely to happen again—but detailed source material of the kind available in the early years of Gambit and Corona had become a casualty of the records destruction process by 1970, so there is no assurance that all of the important events can ever be captured for historians. (RP, March 1973)

Early in August 1967, Martin suggested to Dr. Flax that Gambit systems left over from the program and brief summary records of the achievements of the Gambit program be encapsulated "for eventual release to the Air Force Museum and the Smithsonian Institute." Flax promptly agreed and issued appropriate instructions to the Air Force Chief of Staff and the Commander, Strategic Air Command (in whose facilities the artifacts would be stored). Flax gave the enterprise the name, "Project Van Winkle."[58]

In due time (and rather more time than originally planned), two large sealed canisters went to Vandenberg Air Force Base with instructions for long-term storage under continuing guard. Each contained both hardware and documents, and each carried a plaque explaining that the contents were neither explosive nor toxic but that they could be opened only by approval of the Secretary of the Air Force—at some future time. Flax also insured that "each succeeding commander of Strategic Air Command and the Commander of Vandenberg Air Force Base be briefed on what had been done."[59]

The accompanying summary report, enclosed with the two sets of Gambit hardware, covered the main events of the Gambit-1 program. (The summary had to be completely rewritten early in 1968; as first presented to Martin by General Electric, which had a contract for the whole of the preservation work, it was classic engineering and was nearly unintelligible. Martin had the then unofficial program historian, Robert Perry of Rand, prepare an English version.)

The hardware included everything that went into orbit with Gambit above the interface with the Atlas booster. It was a fitting interment.**

** Comparable Corona hardware was subsequently treated in similar fashion, under CIA auspices, but was on display in a closed area of the National Photographic Interpretation Center in Washington.

Endnotes

1. SAFSP, Gambit Program Summary Report 1960-1967, prep by R. Perry, Sep 67, p 124 (hereafter cited as Gambit Summary).

2. Memo, MGen R. E. Greer to Col. W. G. King, 24 Jul 63, subj: Instructions on Second 206 Bird, SP files.

3. GE, 206 Program Report, Rough Draft, Document No. DIN 500200-34-1, (undated, prep Jul 67), p11-6, (hereafter referred to as GE Report).

4. Memo, Greer to King, 24 Jul 63.

5. Gambit Summary, p 1-25.

6. For details, see Vol V this history, p 112, et seq.

7. For Mission details, see GE Report, p 11-7.

8. Msg 7028, EK to SAFSS, 10 Sep 63.

9. Msg, SAFSP-F-13-9-936, MGen R. E. Greer, Dir/SP, to B. McMillan, DNRO; 13 Sep 63. Msg, SAFSS-1-M- 0196, McMillan to Greer, 17 Sep 63; msg, Greer to McMillan, 18 Sep 63.

10. Msg, 0698, B. McMillan, DNRO, to MGen R. E. Greer, Dir/SP, 24 Oct 63.

11. See Vol V, pp 120-135; see also: MFR, MGen R. E. Greer, Dir/SP, 15 Aug 63, subj: Plans for Ultra-High Resolution Reconnaissance, in SP-3 files; memo, R. Gilpatric, Dep Sec Def, 5 Sep 63, subj: Discussion with Mr. McCone regarding NRO, in NRO Staff files; ltr, A. D. Wheelon, D/Dir (Sat), CIA, to Dr. B. McMillan, DNRO, 5 Nov 63, no subj, NRO Staff files.

12. GE Report, p 11-8.

13. Msg, B. McMillan, DNRO, to MGen R. E. Greer, Dir/SP, 12 Feb 64.

14. Rpt, National Reconnaissance Program Status, 29 Jan 64, SP-3 files; Gambit Summary, pp 4-22 and 4-25.

15. NRP Status, 29 Jan 64, p 3.

16. See Vol IIB, Lanyard, nominally and occasionally called a "Gambit backup," did not pretend to Gambit-class resolutions.

17. GE Report, pp 11-4, 11-9.

18. Ibid; NRP Status, 29 Jan 64.

19. Msg, McMillan to Greer, 12 Feb 64.

20. GE Report, p 11-4.

21. Ibid, pp 5-18 to 5-21.

22. Msgs, STC to SAFSP, 26 Feb 64;GE Report.

23. Msg STC to NPIC, 27 Feb 64; msg, STC to SAFSS, 28 Feb 64; msg, SAFSS to NPIC, 28 Feb 64.

24. See Quarterly Program Review, December 31, 1965 (hereafter cited as QPR) for details of GE Incentive Programs. The scoring system had changed slightly up to that time and was overhauled completely in December.

25. Msg, SAFSP to SAFSS, 18 Feb 64.

26. SAFSP to SAFSS, 3 Mar 64.

27. Msg, BGen R. E. Greer, Dir/SP, to B. McMillan, DNRO, 4 Mar 64; msg, BGen J. L. Martin, Dir/NRO Staff, to Greer, 17 Mar 64.

28. Msg, MGen R. E. Greer, Dir/SP, to BGen J. L. Martin, DNRO Staff, 25 Mar 64.

29. Ibid; msg, NPIC to AF Processing Lab (AFSPPL), 27 Mar 64.

30. Msg, 25 Mar 64.

31. Memo, J. Q. Reber, Chm, COMOR, to DNRO, 26 Mar 64, subj: Request for Epheimeris Data for Next KH-7 Mission, in SS files.

32. Memo, BGen J. L. Martin, Jr, Dir/NRO Staff, to Chm COMOR, 27 Mar 64, subj: Target Priorities, in SAFSS files.

33. Msg, STC to SAFSS, 27 Mar 64.

34. Msg, STC to SAFSS, 16 April 64.

35. Msg, NPIC to AFSPPL, 6 May 64; GE Report, pp 11-12; QPR, 31 May 64.

36. Staff memo for SAFSS-6, (undated, but prepared 3 May 64); msg, SAFSS to SAFSP, 7 May 64; msg, CIA (Col. J. Ledford) to BGen J. L. Martin, Dir/NRO Staff, 1 May 64; msg, DNRO to CIA, 4 May 64.

37. Msg, NPIC to SAFSP, 10 Jun 64; GE Report, pp 11-13; msg, 1,4 STC to SAFSS, 30 Apr 64.

38. Interview, Col W. G. King, Dir/Program 206, by R. Perry, 27 Jul 64.

39. QPR, 30 Sep 64. Many of these details there specified are in conflict with those provided by the GE Report, cited for previous flights. For various reasons, the program review document seems more reliable.

40. GE Report, pp 11-17; The time lag between the 12th and 13th flights was 15 days, which represented a speed-up of only five days from the originally scheduled 28 October launch. Considering the problems created by launch schedule acceleration, however, that represented a considerable achievement.

41. SAFSP, Quarterly Program Review, 31 Dec 64.

42. GE Report, pp 5-6 and 5-7. Such tests began with vehicle 966, used for the 16th flight. A final form of test procedure was introduced with the vehicle for the 26th flight. This combined the operation-during-vibration idea with an accurate mission profile to simulate on-orbit events.

43. Interview, MGen R. E. Greer by R. Perry, 20 Nov 64.

44. GE Report, pp 11-18.

45. Ibid.

46. GE Report, pp 11-4 and 11-19; Gambit Summary, p 3, Attachment 1.

47. QPR, 30 June 65.

48. GE Report, pp 11-20.

49. QPR, 30 Jun 65, Procurement Section.

50. QPR, 30 Sep 65; 31 Dec 65; see also Ch V of Vol V, this mss, and particularly pp 211 et seq.

51. QPR, 31 Dec 65.

52. Memo, BGen J. L. Martin, Dir/SP, to DNRO, 29 Aug 67, subj: Summary Analysis of Program 206 (Gambit), atch 4; Interview, Martin by R. Perry, 8 Aug 67.

53. Gambit Summary, Attachment 1.

54. QPR, Sep 65.

55. QPR, 31 Dec 66.

56. GE Report, pp 11-4, and Gambit Summary, Attachment 1.

57. QPR of 31 Mar 67; 30 Jun 67.

58. Memo, BGen J. L. Martin, Jr, to SAF(R&D) (Dr A. Flax), 4 Aug 67; subj: Long-Term Storage of Gambit Hardware at Vandenberg AFB (Project Van Winkle); memo, Flax to Cmdr, SAC, 25 Aug 67, same subj, all reproduced in Program Summary Report, Vol 1, 6 Mar 68.

59. Memo, BGen J. L. Martin, Jr, Dir/SP, to Dr A. Flax, DNRO, 6 Mar 67, subj: Long-Term Storage of Gambit Hardware at Vandenberg AFB (Project Van Winkle), in SAFSP files; memo, Flax to Cmdr, SAC, 25 Aug 67.

The Development and Operation of Gambit-3

Background and Nomenclature

Gambit was the first operational American satellite system to return high resolution photography.* Originally designed around a lens of 77-inch focal length to produce photographs with ground resolutions of two to three feet, the Gambit was boosted to orbit by an Atlas-Agena combination. The camera system was housed in an orbital control vehicle built by General Electric, an innovation in photosatellite design intended to overcome the assumed stability shortcomings of the Agena. The camera system was a product of Eastman Kodak design; the recovery capsule was adapted from one first developed by General Electric for the Corona satellite. Operational use of the original Gambit system began on 12 July 1963 and continued until 4 June 1967. During that time 38 vehicles were launched. The successor surveillance satellite in the National Reconnaissance Program was Gambit-3.

During its development and operational life, Gambit was identified by several designators other than its code name. "Cue Ball" and Program 206 were respectively a classified non-Byeman cover name and the "white" program designator. Neither was publicly identified with satellite surveillance. Gambit-3 was often called "Gambit-Cubed" although the "-3" designator was actually a suffix differentiating that particular design from three others: Gambit-1, the original, Gambit-2 a proposed modest improvement, and Gambit-4, a proposed very-high-resolution system considered as an alternative to Gambit-3 for development as a second-generation surveillance satellite.

Gambit-3 was influenced by but was not directly related to the Valley system, a very-long-lens development project conducted by Eastman Kodak in the early 1960's. The lens system derived from Valley work was much more closely akin to the proposed Gambit-4 than to Gambit-3. When first considered, Gambit-3 was also informally referred to as Program 207, implying that it was the follow-on to "Program 206," as Advanced Gambit, and G^3—or "G-Cubed." G-3 eventually became the accepted shorthand designator for the successor system, although upon the completion of the original Gambit program and the start of Gambit-3 operations that suffix was dropped.

---

* As noted elsewhere, the abortive Samos E-3, E-5, E-6, and Lanyard systems were intended to perform surveillance functions of one sort or another, but none ever became operational and only Lanyard produced satellite photography.

Old hands continued to use "G-3" as a convenient way of distinguishing the successor system from its predecessor. For reasons of clarity, that distinction has been retained here. In the following sections, the term Gambit-3 is ordinarily used to identify the "advanced" system, and Gambit-1 the original. Unless otherwise indicated when the term Gambit without a suffix appears in quotations, it can be interpreted as referring to Gambit-3, that being the official post-1967 nomenclature. In the Talent-Keyhole designation system used to identify the products of photographic reconnaissance operations, Gambit-1 products were labeled KH-7, and Gambit-3 products KH-8.

The Origins of Gambit-3

Even before the first of the original Gambit reconnaissance satellites had been launched in July 1963, planners acknowledged the need for a more capable surveillance system. By implication, they suggested that such a system could be successfully developed.

The underlying rationale for satellite surveillance stemmed from assumptions that significant operational and technical details of Soviet weaponry could be discovered through satellite photography. Something could be learned from photography with resolutions of two to three feet—which was from three to five times better than anything Corona, the only other available system, might then produce. But much more could be deduced if photographs capable of resolving ground details one foot or less on a side were returned for analysis, and the intelligence community wanted "much more."

The dominant ingredients of higher resolution tended to be focal length (which by implication included optical aperture), and pointing accuracy (which included stability). Smear, a product of imprecise camera stabilization and imperfect image motion compensation, was not always treated as a major constraint in the effort to obtain high resolution. Nor, for that matter, was focal length alone. In the early 1960's the objection to using very-long-lens systems was more a matter of system weight, and principally weight of optics, than magnification potential.

Long lens systems created enlarged images of relatively small areas. In that circumstance, pointing accuracy was essential; surveillance targets had to be caught within the optical field of the lens system if the total system was to be functional. It was customary to design lens or mirror motion into camera systems to cover a wide swath of the earth. But panoramic coverage at high magnifications required huge film quantities, and given the relatively limited film capacity of the boosters and orbital vehicles available in the early 1960's, using panoramic coverage techniques as a substitute for pointing accuracy was not an attractive option.

Lacking a better solution, camera designers of the early 1960's had to adopt a "brute-force" gross coverage approach—essentially a "broad swath" technique. That was the essence of the early Valley program, a proposal to carry large quantities of high-acuity film to compensate for what was assumed to be an inherent deficiency in camera pointing accuracy. Pointing with the precision required to operate a narrow-swath camera system was until 1963 generally assumed to be beyond the capability of satellite-carried cameras.

Recovery of the first Gambit-1 film in July 1963 disclosed that several system capabilities about which there had been significant doubt were feasible. Although few photographs were recovered from the first Gambit mission (because Major General Robert E. Greer was determined to obtain "one good picture" and not to endanger that goal by attempting complete system operation on the trial mission), the consequences were enormous. Getting the pictures was one substantial achievement;* the obtained resolution of about 3.5 feet was another; and exceptional pointing accuracy was a third. Although the first two were of greater immediate importance, the demonstrated ability of Gambit-1 to point its optical system with great precision caused a complete revision of long-term plans for a next-generation surveillance system. Valley was promptly redirected toward "narrow swath" concepts; the "wide swath" and "brute force" approaches were abandoned as unnecessary. By August 1963, Valley research and Gambit-1 experience had convinced the National Reconnaissance Office that long focal lengths were feasible for satellite operations and that (because pointing accuracy could be guaranteed) a system built around large optics could be appreciably lighter than had earlier been thought necessary. One consequence was that plans for the "big optics" system were adjusted to provide for use of an Atlas rather than the much larger Titan III-C booster (although other considerations later caused a reversion to the Titan booster).[1]

Disagreements and uncertainties marked subsequent developments. A major contributor was a bureaucratic competition for control of the satellite reconnaissance program. But for the most part such skirmishing concerned matters other than the "very high resolution" system, which in July 1963 received the endorsement of Dr. E. M. Purcell and a special panel he headed** and in September became the subject of preliminary plans for contractual actions.[2] One of the problems was technical: the Purcell Panel had recommended development of specific lens systems defined in terms of focal length and aperture (and very large aperture at that—f/2.0 and f/1.5 lenses of 60- and 40-inch diameter were proposed!). General Greer was convinced that the proper course was to require a specific ground resolution and let the system design emerge from the tradeoffs necessary to obtain that resolution.[3] Ultimately that became the approved course of Gambit-3 development, but in 1963 it was not popular.

Although several contractors had been involved in the "big optics" program of 1961-63, Eastman Kodak had two notable advantages when proposal time arrived, having been the principal contractor for Valley and having designed the original Gambit system. It would have been feasible, of course, to have some contractor not involved in either Valley or Gambit develop a Gambit successor, but cost, time, and technical capability factors all agreed against such a course.

Early in December 1963, Eastman Kodak presented ideas and concepts for an advanced Gambit system to Dr. Brockway McMillan*** and (subsequently) to General Greer. As a result of the presentation, Greer formally proposed the development of a higher resolution, longer lens Gambit system.

The crux of Eastman's proposal was to develop a system that would exploit the pointing accuracy of Gambit-1 to drive a camera with a length lens. Eastman urged that resolutions could be obtained from such a system, assuming photography from an orbital altitude, Eastman also proposed that the new system incorporate a "factory to pad" concept providing greater modularity than Gambit-1.

---

* Apart from Corona, which had been operational for three years, one Lanyard flight of May 1963 which produced a few photographs of no great intelligence worth and the returns from one Samos E-1 mission (with resolution limited to about 100 feet), represented the only previous successes of a satellite reconnaissance effort that had been in existence for nine years and had been heavily funded for five. Corona, sponsored by the CIA, was not considered an element of the "Air Force" satellite reconnaissance program, being classified as an "interim" capability system even though developed, managed, and operated mostly by Air Force people. Both the Samos E-5 and Samos E-6 programs had failed and had been cancelled by the end of 1962—after eight consecutive mission failures (nine, if the first Lanyard were counted). An effort that very probably cost more than a IV had yet to produce useful photography. Greer's concern for "one good picture" was all too understandable in those circumstances.

** The Purcell Panel simultaneously urged discontinuance of efforts to develop new optics for Corona and in so doing provided the initial impetus for the S-2 and Fulcrum programs. By 1966, the Hexagon program had emerged, laboriously, from Fulcrum and S-2 activities. See Chapter XIII.

*** Director of the National Reconnaissance Office from February 1963 to October 1965.

Specifically, instead of an orbital control vehicle which enveloped the camera system, Eastman proposed using two modules, physically distinct, one containing the camera and the other providing orbital control. That arrangement would make it unnecessary to do major assembly work at the launch complex, a troublesome aspect of Gambit-1 operations.[4]

Although the proposed Advanced Gambit would include a variety of detailed refinements of the basic Gambit system, Eastman Kodak felt that the major system performance improvements would derive from three factors: focal length, the use of International Harvester's mirror substrate, and introduction of a new thin-base, high-resolution film with a substantially higher exposure index than that available for Gambit-1.*

Three candidate systems had received attention during the 1962-1963 period when a higher-resolution successor to Gambit-1 was under consideration. Categorized in terms of resolution potential,"Gambit-3" and "Gambit-4" would have a much better resolution potential than the "Gambit-1" and "Gambit-2." (In each instance the requirement postulated 95 percent returns at the desired resolution.)

Gambit-2 did not appear to afford a sufficient advantage over Gambit-1 (which had a potential for resolution for rather less than 60 percent of its photography). Development of Gambit-4 promised to be attended by "...serious technological and manufacturing uncertainties, formidable costs, and a long development time." In the view of the National Reconnaissance Office (NRO), Gambit-3 would provide "...significant improvement in resolution at acceptable estimated costs and lead times, " and should be the preferred system. Eastman Kodak heartily concurred, although disagreeing with the additional estimate of NRO program officials that the Gambit-3 program might incur major problems if the new mirror substrate material and Eastman's higher speed film did not become simultaneously available. To hedge against any major program difficulty that might arise from that circumstance, the NRO decided to sponsor the concurrent development of the primary substrate material endorsed by Eastman and of two alternatives.

By 13 December 1963, Colonel W. G. King, Project Director for Gambit, had constructed an initial development proposal based on technical content of Eastman's "Gambit-3" briefing. He incorporated Greer's instructions that general cleanup of the Gambit-1 system would continue until Gambit-3 became operational.[5]

King's plan provided that the entire Advanced Gambit program would operate under the purview of the existing Gambit office. The initial flight of the new system was scheduled for the second quarter of 1966, with the operational transition from old to new system taking place later the same year. Contractors for the new system would begin "controlled entry" into development toward the end of fiscal 1964.

King saw only two feasible options for orbital control. He conceded that an orbital control vehicle could be developed with a capability similar to that of Gambit-1, but for various reasons he favored using an Agena with a roll joint coupling to the photographic payload. The roll joint used in the first few Gambit-1 flights had operated perfectly, while General Electric's orbital control vehicle, a new and complex system, had experienced a great many difficulties.** To rely on the Agena for orbital control afforded advantages of lesser technological and financial risk.*** King's tentative schedule called for receipt of proposals in mid-April of 1964 and contract award by June.[6] As 1963 ended, McMillan informally authorized Greer to proceed with the Gambit-3 program outlined in King's development plan. Formal approval appeared on 3 January 1964.

King and Greer had been predisposed toward the combination of Atlas and Agena, but McMillan felt that as long as scheduling difficulties would not result, the option of using a Titan III should be retained. That change represented the only significant initial alteration of King's Gambit-3 plan.[7]

King and Greer worked out the remaining major elements of the Advanced Gambit concept early in January 1964. The major subsystems of the satellite were to include a photographic payload section (PPS), a satellite control section (SCS), and the booster vehicle. The payload section would contain a camera module and a recovery vehicle (SRV). The control section was to include the orbit adjust module, an attitude control subsystem, a back-up stabilization system (BUSS), and the power supply.****

---

* Film with an exposure index (EI) of six might not seem "fast" to commercial users, who in 1963 were able to buy EI 1200 film from Kodak, but EI six film was roughly three times as fast as the film used in Gambit-1. Its availability made the lens-mirror system of Gambit-3 an equivalent two f-stops "faster" than the predecessor system.

** The roll-joint had been invented for Lanyard and later adapted to Gambit-1.
*** Although experience with the GE orbital control vehicle was limited in early 1964, to Colonel King it was evident that the subsystem was potentially the most troublesome of all Gambit elements. The accuracy of that assessment was evident two years later: General Electric proved difficult to work with, had the poorest cost record of any Gambit contractor, had the poorest schedule record, and delivered systems with the poorer reliability than that of any major Gambit subsystem.)
**** Although the option of using a Gambit-I-style orbital control vehicle had not yet been formally discarded, the Greer-King concept was heavily biased toward Eastman's proposed approach as early as January 1964.[8]

A preliminary development plan for what was to become Gambit-3 appeared early in February. It made clear the contributions of Gambit-1 to the early concept of what was already being called "Gambit Cubed."* Operational use of Gambit-1 was then demonstrating that testing and maintenance were greatly hampered by the highly integrated nature of the subsystems. Thus, said program managers, "...a great deal of connecting and disconnecting of wiring harnesses was necessary...In the G^3 system, the emphasis on modular construction will increase the ease of maintainability.... The command control subsystem... of the G-Program has been shown to be a superior system and it is anticipated that it and associated software may be included in the G^3 system." Thus, not only would the direction of Gambit-3 development be guided by Gambit-1 experience, but some of the hardware and software of the earlier system would be used "off the shelf" for the new system.[9]

Colonel King expected to have program evaluations and recommendations complete by 13 May 1964, to have all contracts in effect by 30 June 1964, and to meet a scheduled first flight date of 1 July 1966.[10] In the event he was optimistic only by four weeks.

Only one major change to the preliminary specifications occurred in the early stages of development. After consideration of the risks, General Greer decided it would be imprudent to anticipate full development of a dual platen camera configuration in time to meet flight schedules. McMillan agreed. With that change, negotiations with Eastman Kodak for the development of the camera system could be concluded. Contract price, including fee, was just under.[11]

By May, Eastman had completed preliminary research that seemed to reaffirm the feasibility of the primary mirror and had concluded that use of the new thin base, fast film would allow three times as many photographs to be taken on each Gambit-3 mission as on Gambit-1 missions...if problems of storage, on-orbit life, and recovery of larger film loads could be solved.[12]

Neither a booster nor an orbital control system had yet been chosen. Cost and schedule implication were large. Moving from Atlas to Titan III would imply either building a new pad at Point Arguello, or transforming an existing Atlas pad. Building an orbital control vehicle that could initially be used on an Atlas and later on Titan was a complex operation. If Titan were chosen, with its much greater lift capability, useful compromises in camera system weight and performance might be feasible.[13]

Predictably, Lockheed favored use of the Agena and GE the use of an orbital control vehicle (OCV) like that of Gambit-1. Both contractors had to consider the adaptability of their vehicles to the Titan III booster.

Influenced by orbital control system problems which were already beginning to affect the Gambit-1 flight program, Greer and King had early endorsed the use of Agena, suitably modified, as the orbital control vehicle, although conceding the need for backup work by GE to help keep all options open.[14] (During the first few Gambit-1 missions, the orbital control vehicle was only used in "solo" flight, after the film had been recovered. Primary missions had been flown in what was known as the "hitch-up" mode, which meant reliance on the Agena for orbital control by way of a roll joint. Results had been greatly better than expectations.)

The Agena was a proven space vehicle. Its shortcomings were well known and generally were not critical. The orbital control vehicle developed by GE both duplicated Agena capabilities and provided desirable properties the Agena did not possess. Lockheed proposed extending proven Agena capabilities to meet the requirements of the Gambit-3 system. GE proposed modifying the existing Gambit-1 orbital control vehicle. Greer, concluding that the OCV approach was too complex and too risky recommended that Lockheed do the primary vehicle study and GE a backup study. McMillan agreed.[15]

Work on the parallel satellite control section studies began in early July. Lockheed's task was to study the compatibility of the Agena with the Titan III-X, the version of Titan best suited to the Gambit-3 mission. GE was encouraged to look at both the simplification of the GE proposal for a simplified OCV and the possibility of using a separate ascent stage (rather than Agena) with the orbital control vehicle.

The photographic payload section included three major components, the satellite recovery vehicle (SRV), the external structure and the camera itself. Eastman had originally agreed to build the recovery vehicle thus concentrating the design, development and manufacture of the entire payload section in the hands of one contractor. It rapidly became clear, however, that Eastman would have difficulty in staying on schedule with just the camera subsystem to worry about. In consequence, Eastman subcontracted the capsule work to the Recovery Systems Division of GE, which had built most earlier recovery capsules. Similarly, Eastman subcontracted external structural work to Lockheed and concentrated on the development of the advanced optical system for the new Gambit-3. However, the arrangement that made Lockheed and GE subcontractors to EK insured that

---

* Why "G^3" ("Gambit Cubed" or "G-Cube") became the informal designator for the program, rather than "G-3" ("G-Three"), is a minor mystery.

both the external structure and the reentry vehicle would be designed in compliance with camera system needs and that design compromises, should they prove necessary, would be compliant to the primary needs of the camera system.

By the Fall of 1964, Eastman and its subcontractors had advanced to the point where engineering specifications of the composite design had been released, design reviews held, some engineering drawings had been released and some of the critical long lead-time items had been ordered. Nevertheless, payload development was somewhat behind schedule.[16]

The hedge represented by concurrent development of two orbital control concepts was duplicated, by force of circumstances, in camera system development. Neither King nor Greer was willing to hinge program success on the probability that both the new film and the new substrate material would appear, operationally ready, on schedule. The two-fold task of casting such large blanks as were required for the stereo mirror and for the primary mirror) and then polishing them to the required smoothness provided numberless opportunities for delay. Eastman had originally estimated that each of the two mirrors would require almost 800 hours of grinding, polishing, in-process testing, electroplating, and coating. The actual manufacturing time proved to be about 3000 hours for each mirror blank, thus the precautionary development of the two alternative substrate materials. As substrate development proceeded, initial pessimism proved to be warranted.

The program experienced several failures; collapsed and fracturing substrates, and polishing processes which proceeded far more slowly and less accurately than expected. The results were so variable that no final selection of a substrate material could be made in 1964.

During the last quarter of 1964, Eastman fell farther behind schedule while all other phases of the Gambit-3 project were either on or ahead of schedule. Less than a year old, and therefore going through a period normally associated with legions of bad surprises, the Gambit-3 program had encountered and overcome most of its major technical problems. Only the camera optics still presented significant difficulties.

Despite such rapid progress, by the end of the calendar year the Special Projects Directorate was able to return unexpected Gambit-3 funds to the NRO budget, a precedent that was continued, by and large, for another eight years.*

---
* Through 1972.

Although schedules continued to slip, by late 1964 Eastman had apparently solved the problem of manufacturing the optics for G-3. The solution lay in foregoing unconventional manufacturing and polishing techniques for conventional ones. The reason for going to unconventional techniques in the first place had been the difficulty of precision finishing mirror surfaces. Two developments of 1964 made it possible to overcome the natural shortcomings of the conventional techniques: mapping the surface irregularities of the mirrors by interferometery, and filling the mapped depressions by electro-plating. It had taken Eastman more than a month to prove the applicability of these processes however, and that represented a month of schedule slippage.[17]

Other equipment to be included in the photographic payload was subject to changes. The Astro-Position Terrain Camera, completed early, was redesigned by the end of the year to enhance its performance. GE completed design of the satellite recovery vehicle by November 1964. Profiting from earlier experience, King and Greer had insured that the G-3 recovery vehicle was much like that of G-1. The only significant change from the original was the imposition of extremely stringent quality control processes from the start. The widening schedule gap in optics development thus was the principal subject of management attention. New controls, technical reviews by Aerospace Corporation personnel, and the institution of bi-monthly progress reports were all aimed at getting Eastman back on schedule, or failing that, to prevent further slippage.

In October 1964 on the basis of the Titan III X-Agena studies carried out by Lockheed and General Electric, Greer's staff prepared cost estimates pertaining to a switch from Atlas-Agena.** Just before the end of the year, McMillan approved the change to Titan.[18]

That decision had a number of effects on the program. The first was a significant growth in the budget. Another was a slippage in the scheduled initial launch data. But from the long view, neither of these perturbations was important. The choice of a Titan booster made future system improvement far simpler than it would have been if the lift capacity of the Atlas had been an operative constraint.

The collective experience of McMillan, Greer, and King, in particular their recent Gambit experience, induced a sequence and timing of Gambit-3 decisions which, in retrospect, may have assured the ultimate

---
** A consideration of some importance was the probability that a new search system to replace Corona would have to rely on Titan III boosters. Although considerable technical and institutional controversy preceded the latter (April 66) decision to develop what became Hexagon, the pattern was plain enough for Greer to see in 1964.

success of the Gambit-3 program. Ironically, the keynote of that decision process was indecision—or delayed decision. They refused to allow the program to be tied irrevocably to the "obvious" booster (Atlas), the "available" orbital control vehicle, or the "most promising" mirror substrate material. In each area, they delayed design freezes long enough to allow all the relevant data to come to light. The price of such decision delays was a growth in program cost and a modest slippage in the initial launch date. The product seemed well worth that slight cost.

The second year of Gambit-3 development saw the disappearance of residual ambiguities and uncertainties affecting the final configuration. In particular, optical materials and manufacturing processes were selected and adaptation to a Titan III booster mode was completed. By the end of the year, engineering models of all systems had been built and were in test or fabrication stages. The transformation of Point Arguello Launch Complex II (PALC II), Pad 3, from Atlas-Agena capability to Titan III-Agena capability was well underway. (An additional capability of handling small, strap-on solid rocket motors had been added.)

The distinguishing marks of the Gambit-3 program were that the problem-solution cycle was more compressed than in earlier programs of similar nature and the program experienced continuing economies rather than overruns. Once the schedule adjustment required by the shift to the Titan III booster had been accommodated, the schedule remained valid. The projected capabilities of Gambit-3 equipment were enhanced rather than reduced as the program proceeded. It is interesting to recall that such events took place during a time in the mid-1960's when virtually all military procurements were experiencing major cost overruns, schedule slippages, and performance shortfalls.

By the beginning of 1965, major camera system problems had been reduced to four. One was the selection of a substrate material (and a structural form) for the primary and secondary mirrors. The requirements for the material included low weight, dimensional stability, and handling properties appropriate to the method of surface preparation. A second problem, partly solved, was the development of a method for preparing the precise optical surfaces. The requirements here were that the method yield the desired precision, but that it do so both economically and at a rate of production consistent with the operational plans for the completed satellite. The third problem was to devise a method of mounting the optical surfaces in the spacecraft so as to maintain focus and avoid the introduction of mechanical distortion. Finally, some technique had to be developed by which thermal distortions could be controlled during operations.

Program reviews during the last half of 1964 had dealt chiefly with the first two problems. One result of the concern evidenced by those responsible for reviews was to amend the Eastman Kodak contracts to include reporting on critical experiments and tests which had a bearing on the issues of concern. This activity came to be called the "Early Demonstration" program. After reviewing it during September of 1964, McMillan directed that the program be extended to include all aspects of the optics development program. He anticipated final selection of substrate materials by May, final decisions on engineering by June, completion of the engineering by August and completion of tests during September.[19]

Concentration of attention and effort on one component of one subsystem of the project had some interesting implications. Beyond the critical nature of that component, and the fact that it represented the largest single technological advance captured in the Gambit development, the attention paid it suggests that there was little else of critical moment in Gambit-3 development that required much management attention. In any case, management spent its time rather lavishly on the one major problem that did exist.

Early in January 1965, notification of the switch to Titan III Agena configuration went to the principal contractors. It was the last program change of any consequence in Gambit-3 development.[20]

Unhappily for routine operations, the Titan III announcement very nearly coincided with a major program review, having the effect of invalidating most of the schedules on which that review was predicated. As it happened, however, except for the mirror development activity the entire program was essentially on schedule—and within predicted cost ceilings—and the readjustment necessitated by shifting to a Titan III launch capability provided a cushion for the mirror development schedule.[21] Formal direction for the incorporation of the new booster in the Gambit-3 program included instructions that the budget was to be austere, pad modification was to begin at once at PALC II, production rate of the boosters was to be approximately one per month, and the first should be available in time for the projected first launch date of 1 July 1966. The project staff were also directed to include a capability of using small strap-on solids (such as the Minuteman first stage) in both the pad modifications and the booster production contract.[22]

The optics situation was still sufficiently worrisome in March 1965 to warrant a four-hour briefing for Dr. McMillan.[23] Many substrate materials still were being considered.[24]

By early March, Martin-Marietta (the builder of the Titan) had concluded that the earliest possible delivery date for the first booster could not support a launch date before 28 July 1966, almost a month later than the date by which the payload and satellite control sections could be ready. There seemed to be no way to protect the original first flight date of 1 July 1966, so General Greer recommended that the late July date be accepted. Dr. McMillan concurred immediately.[25] Both were aware that Eastman's payload schedule slippage made the goal of 1 July more theoretical than real in any case.

March 1965 marked the appearance of a new problem. The satellite control section began to fall behind schedule when GE started work on a new command system. That development paced the progress of the checkout for the development test vehicle, which by the end of March had fallen six weeks behind schedule. In the end, therefore, the new booster schedule caused no real problems.[26]

The progress of Gambit-1 and its flight experience were also important for Gambit-3. During the Spring of 1965, the project office had decided that Gambit-3 should incorporate both a redundant view port actuator and a backup film cutter. The camera-door actuator certainly owed its provinance to a series of failures experienced on Gambit-1 flights. The Gambit-1 systems, primarily pneumatic, continued to fail throughout the history of that vehicle until, near the end, they were completely replaced by electro-mechanical devices, While the lesson might have been that pneumatic devices could not be relied upon for space operations, it was not accepted in total at the time. But more of the primary actuators used on Gambit-3 were electro-mechanical than had been the case for Gambit-1.

By the end of June 1965, GE was two months behind schedule in command system engineering, and Eastman was a full three months behind in mirror development and fabrication. Happily, neither delay impacted on the initial launch date schedule, for which the modification and structural checkout of the new Titan III was the pacing item. In addition, test schedules for virtually all subsystems of the Gambit-3 were very liberal, so slippages in development simply meant some low risk compression of the test programs.[27]

The central problem at Eastman involved more than simple inability to solve the manufacturing and finishing problems for optical surfaces. As the project office phrased it in June 1965, "It is evident at this time, that EKC has underestimated the engineering manpower required to produce the electromechanical portion of the PPS (photographic payload section). Their late design releases are evidence of this. We have started down too many paths in some areas and have been late in formalizing the route to be taken."[28]

Extreme concern for preserving options had earlier in 1965 been expressed in the wide variety of substrate materials which continued under investigation. By June, indecision had ended. Eastman still followed a course of caution, however, keeping the new alternative material, as an open option. In the event, the first attempt to produce a second generation failed, and the backup material, fused silica, was elevated to primary status. *(Paragraph and Associated Endnote (29) Redacted.)*

Perhaps a more significant change occurred even before the shift to fused silica. The original stereo mirror had a tapered design, which caused trouble in the polishing process, and its beveled edges were troublesome to grind. Both these special features were eliminated with a new design.[30] Although the simplification cost a slight loss of theoretical resolution, that was never to prove a serious drawback.

During the subsequent nine months, new authorizations, cutbacks, and transfer of funds to the fiscal 1966 budget raised the actual figure to only less than the original estimate. The ability to predict costs so accurately—for a high-risk development program—was perhaps unique to the Special Projects Directorate among all major government contracting bodies of that era. *(Paragraph and Associated Endnote (31) Redacted.)*

During the summer of 1965, the NRO experienced administrative changes which, while of considerable internal significance, had only minor influences on the Gambit-3 program. After some two years of struggling with the CIA over control and direction of the National Reconnaissance Program, Dr. McMillan left his post as Director of the NRO in September. He was replaced by Dr. A. H. Flax, Assistant Secretary of the Air Force (R&D).* General Greer, with the successful development and early operations of the original Gambit to his credit, retired from the Air Force, to be replaced as Director of Special Projects by Brigadier General J. L. Martin, Jr. Martin, in turn, was succeeded as head of the NRO staff by Brigadier General J. T. Stewart. Colonel King remained as project director for both Gambit-1 and Gambit-3, thus insuring management continuity.

Schedule problems continued to trouble Eastman, General Electric, and Lockheed. Photographic payload section schedules were so tight that for the first time the project office began to question whether

---

* Flax served as Acting Director at various times between July and September 1965, during McMillan's temporary absences. McMillan's plans were known to the NRO staff in July.

the initial launch schedule could be protected: "Continued compression of the EK test schedule is beginning to make the 28 July 66 initial launch date appear somewhat optimistic," King's group noted in September. Similar doubts persisted through the end of the year. Eastman was six months behind by September 1965, and the reliability model test sequence, originally scheduled to begin on 15 October 1965, had to be rescheduled for April 1966.[32]

Although pad modification at Vandenberg Air Force Base was 70 percent complete by the end of September, labor jurisdiction problems briefly caused worry about a work stoppage. Apprehensions had two facets. Not only would the launch schedule be invalidated by a stoppage, but there was no desire in the project office to call attention to Gambit-3 activities at Point Arguello.

Minor difficulties also arose in Lockheed's effort to incorporate new redundancies into the control electronics for the roll joint while Martin encountered combustion instability symptoms in Titan III-X tests.

Eastman's difficulties had several sources, not the least being an apparent overcommitment of resources. The company was working simultaneously on Gambit-1, Gambit-3, a lunar camera for NASA, and a proposed new search system (for what later became the Hexagon competition). Late in 1965, Eastman also began to do work on mission profiles for Gambit-3. (Such profiles would simulate the flight of the camera around the globe and yield the data necessary to study on-orbit camera operations. The data would then be used to derive operational parameters for the camera during a variety of flights. The parameters to be estimated included the number of roll maneuvers, distribution of roll angles and the distribution of slit positions. Because of the problems encountered on Gambit flights with the horizon sensors, separate profiles had to be developed for winter and summer operations.)*

Even beyond the administrative changes and schedule slippages encountered with Gambit-3, the summer of 1965 was rather traumatic at the West Coast facility. Gambit-1 problems had come to a head during July and August: three successive catastrophic failures caused a complete stoppage in the receipt of high resolution photography. The immediate result, felt during the fall of 1964, was an intensification of virtually all quality control measures imposed on National Reconnaissance Program contractors. The first Gambit-3 item affected by such stringency was the Lockheed command system; after some difficulty, the project office brought Lockheed to agree to the inspection of all command system components, beginning with the second batch of equipment (to be used on flights after number five). More stringent and earlier tests were also required for the first group of flight items as well. Finally, Lockheed was encouraged, in some instances, to find better qualified personnel to work on Gambit-3.[34]

Contracts in virtually all areas had been let for development and procurement of hardware for the first six (developmental) flights.[35] After exposure to the flurry of Gambit-1 problems, General Martin concluded that requirements for more intensive inspection would not, of themselves, instill a sufficient degree of system reliability in either Gambit. He concluded that the procurement policies hitherto used for the program were best suited to a developmental program relatively low in technical and high in financial risk. The environment in which Gambit had been developed explained adequately the kind of incentive structure which could be thus described. The Gambit program was no longer in a development phase, however, and something new was needed.

Enough experience had been gained with Gambit production to give the project office considerable confidence in its ability to estimate the cost of a satellite. But the Gambit-1 failures of the summer of 1965 indicated that performance was far less predictable. In particular, existing contractual incentives for saving money in production were greater by a factor of about two than those for performance. Appreciating such considerations, General Martin developed a new incentive system which not only reversed the order of priority, emphasizing performance over cost, but deleted all reward for cost saving. That policy was consistent with Martin's belief that the cost of the hardware was the least of existing uncertainties; he perceived on-orbit performance to be the crux of the problem. That Gambit contractors accepted the new incentive structure strongly suggested that the new

---

* See chapter on Gambit Flight Operations. During deep winter, the heat sensitive horizon sensor aboard Gambit was incapable of distinguishing between space and earth over the South Pole. After a brief flirtation with the development of more sensitive devices, a solution rather typical of the Special Projects group was found; since photographs of the polar ice cap were not of interest, neither was orbital stability in that region. The vehicle was therefore allowed to coast over the cold region, regaining attitude reference and control as it reached the areas of intelligence interest. A supplementary benefit arising from that solution was that orbital control gas was expended less rapidly since it was only used on part of each orbit. The expedient was one of several that allowed the eventual stretch out of Gambit on-orbit life from an originally specified five days to eight days by the end of the program.[33]

system made more sense than the old. In fact, the new incentive structure subdued several contract disputes that had been continuing for several months.*

The Martin reforms could scarcely have been timed better for Gambit-3. Program managers planned six development flights of the new system, after which it was to be considered operational. Contracts covering the initial flight systems were converted to the new incentive structure shortly before the first flight of Gambit-3. That development, like the accelerated inspection, checkout and test program for Gambit-3, was a direct product of Gambit-1 experience.**

By the end of 1965 it was apparent that the occurrence of any major problem in the test cycle for the payload section would cause a slippage of the initial launch date for Gambit-3. Eastman was by then 30 weeks behind schedule; it was the low point for the company and the program. Indications of improvement, later shown to be accurate, were present even then, however. Several components of the photographic payload section had completed their initial tests with slight difficulty, an event that brought a symbolic sigh of relief from apprehensive program managers.

The selection of fused silica, the primary mirror substrate, ended uncertainty about initial optics design. Fused silica would be used in the first 22 vehicles (the developmental flights and the first 16 operational systems).*** And by the end of the year construction work on the modified launch complex at Point Arguello had been completed, as had the first battery of tests of the Development Test Vehicle (the satellite control section) at Lockheed's Systems Test Complex. On 6 December, the test article was returned to Eastman for mandatory re-manufacturing to correct faults discovered in testing. At the end of 1965, optimism was in order.[36]

One product of photographic payload section difficulties was an early 1966 reorganization of the entire test schedules for both Eastman Kodak and Lockheed. If schedules could be maintained, and no major test failures occurred thereafter, the new system would make its planned first launch date. But all was not completely serene. By January 1966 there existed considerable doubt that the high-speed (E.I.6), high-resolution film on which Gambit-3 excellence so heavily depended would be ready for use in initial flights. If it were not ready, Gambit-3 photography would not satisfy expectations. The fall-back film, with proper resolution and a lower exposure index,**** would cause a certain amount of smear in the photography; control parameters had all been designed to accommodate the faster film.[37]

Other elements of the development program were going well. Pad modifications still were on schedule. The development test vehicle, having been through mandatory re-manufacturing, successfully completed its thermal-vacuum tests in late March. The integrated satellite control section and photographic payload sections were expected to enter joint testing at Lockheed in early April, as planned. Engineers identified and solved a venting problem with the Titan booster and by March had completed structural tests on its skirt and the Agena adapter. The troublesome command system hardware for the development test vehicle was back on schedule. Lockheed had understood the message implied by incentive and test procedure changes, and was responsive to program needs. Most hardware schedules looked sound. The only significant problem, other than film, was with substrates for the mirrors, and it was now a manufacturing problem rather than a design problem. Casting difficulties persisted; other problems seemed to be under control.[38]

---

* The "Martinized" incentive structure was described in his 1969 paper, "A Specialized Incentive Structure for Satellite Projects." It discussed, clearly and in detail, the thesis and its application, although (of course) Gambit was not identified as the program to which the policy was first applied. General Martin's approach was intended for application to mature programs, those past the stage of schedule and cost uncertainties but still subject to performance improvement. Its attractiveness, and the obvious success of its application to Gambit, had the unfortunate effect of inducing others to attempt to apply it to programs that lacked the fundamental character General Martin specified. In particular, it was occasionally applied, in part or in whole, to the development phases of new programs—where the structure devised by Generals Greer and King for the initial stages of Gambit was almost certainly more suitable. Misapplication of the Martin strategy to immature programs was particularly unfortunate because what may have been General Martin's major contribution was the demonstration that the incentive structure could be "fine tuned" to the needs of almost any procurement problem. He did not intend, and certainly did not recommend, that it be applied to development enterprises characterized by cost, schedule, and technical uncertainties of real consequence. The success of the Martin approach resulted from his recognition that Gambit had changed as it matured, and that a mature Gambit required contract incentives different from those of developmental Gambit. He did not suggest, and did not believe, that the incentives of the original Gambit contracts were in any sense incorrect—merely that they were no longer appropriate to the circumstances he had to contend with. The incentives he devised could, or may have been, of indifferent quality, in an absolute sense. But they were better suited to the circumstances than those they replaced. There lay the explanation of the success of General Martin's approach. And there lay the seed of failure when his techniques were applied, in inappropriate circumstances, to other programs.

** Since Gambit-3 was still in the "development" stage, one might wonder if this did not constitute the first misapplication of Martin's new incentive system. The answer is no for two reasons. First, cost uncertainties were not substantial, nor were technical risks. Second, Gambit-3, as the name implies, was a new mark of a proven system.

*** In fact, although hope continued for the later introduction of the new material, only one mirror made from that substance would ultimately be flown.)

**** The alternative was a film with index of 3.6 with a resolution capability of 110 lines per millimeter, as against the nominal 130 lines ASA 6.0 film under Gambit-3 flight conditions.

Gambit-3 introduced one other significant change derived immediately from experience of the early phases of Gambit-I. In lieu of the extensive testing at the launch site that characterized Gambit-1, testing that frequently was accompanied by substantial amounts of repair work in the Missile Assembly Building, Gambit-3 incorporated a command system with features permitting automated checkout of virtually all vehicle functions, telemetry readout of the functional check being fed directly to a computerized evaluation and assessment subsystem that indicated directly whether or not various subsystems and components were operating within acceptable limits. The automated checkout normally was performed during final assembly of the payload at the principal manufacturing points (EK and Lockheed-Sunnyvale); vehicles, therefore, went directly from factory to launching pad, bypassing one of the most fertile sources of subsequent trouble. The independent subsystem checkout routines were combined into a single simulated flight operation during final check of the vehicle while on the launch pad immediately before launch. The effect of the new capability, and the procedures that accompanied it, was to increase confidence in the validity of the testing process, to abbreviate the launch and countdown procedure, and to eliminate the field handling and testing that had sometimes contributed to later operational problems.[39]

Development Flights

At the end of June 1966, the project officer reported "...reasonably good (indications) that the initial Gambit-3 launch date of 28 July 1966 can be met." The Titan booster had arrived at Vandenberg on 7 May and was mated with the satellite control section of the development test vehicle three days later. Inspectors accepted the control section and command system by 30 June, but refused to sign off on the Titan because of residual thrust instability. (Two alternative fixes were put in train, both of which would protect the launch date.) Final acceptance testing of the photographic payload section went slowly, but there were no failures. *(Paragraph and Associated Endnote (40) Redacted.)*

By 28 July the vehicle was ready to go. General Martin decided to ignore several minor defects in order to bring off the first flight on schedule. The most important, discovered during the pad checkout of the primary camera, was that commands to change the slit size were only intermittently obeyed. The project office decided to fix the aperture in the best "average" position to insure that research objectives could be satisfied. Operation of the roll joint was constrained to limits of plus or minus 35 degrees, a precaution reminiscent of the initial flight of Gambit-1.

At the last moment, an anomaly developed in ground station equipment which forced a one-day delay in launching, but at noon on 29 July 1966, the teletype at the Special Projects office in El Segundo began to rattle off its message: "29 Jul, 1830:22Z Preliminary TLM data indicates normal launch." Precisely two hours later Sunnyvale reported, "All systems appear normal." Target count data began to flow in four days later.[41]

The ephemeris achieved was very close to that sought: inclination of 94.15 degrees, apogee of 150.33 nautical miles and perigee of 84.43 nautical miles. The primary (i.e., photographic) mission lasted for five days, during which the system was programmed for maximum targets. Of these, a total were successfully "read out."* The best resolution actually obtained in the recovered photography was as predicted, measured at a contrast ratio. The film was also underexposed.[42]

Aside from the constraints imposed on vehicle operations before the flight, the only operational failure was intermittent operation of the Astro-Position Terrain Camera. The reentry vehicle was successfully recovered on the 83rd revolution, after which three days of solo flight were conducted with the satellite control section. Shortly after separation of the recovery vehicle, the roll joint malfunctioned (on orbit 90), but it later recovered. The solo flight was used to gain experience with the vehicle in orbital maneuvers and to carry out some nineteen experiments related to the demonstration of specified capabilities of the vehicle. Three successful orbital adjustments on orbits 89, 97, and 122 satisfied the first of those objectives, and all of the subsequent experiments were successful.[43]

---

* To be read out, a photograph must represent the confluence of several aspects of success. The camera must have operated correctly, causing no perturbations which would make the negative unreadable. Such perturbations could include uncompensated smear, incorrect focus, faulty compensation for thermal effects, solar or terrestrial flare and (occasionally) degraded optics. The film had to be unmarred either by faulty manufacturing or by scratches caused by film transport or take up. All of these factors were nominally controllable in manufacture and checkout. The major cause for unreadable film, however, was natural and uncontrollable—weather. The dominant cause for differences between targets programmed and targets readout in the entire Gambit-3 program was cloud cover. In later years, the output of weather satellites lessened that effect, but it would persist as long as cloud cover data were other than instantaneous.

There is almost no stable relationship among frames exposed, film used, and targets covered. Most targets occur in clusters of random size. Thus, several are scheduled to be covered in one photograph. In addition, a "photograph" may be a single frame, a stereo pair, or a strip of variable length, so there could be no predictable relationship between film length and the number of exposures, or between the number of exposures and the number of targets photographed—or between the number photographed and the number "read out."

Post flight analysis of recovered film was revealing. The overall quality of the imagery from the first Gambit-3 mission proved to be better than that obtained from any Gambit-1 mission.*

Eastman analysis noted, "Although...primary optics fell short of the design goal, the intelligence provided by this mission was reported to be the highest of any reconnaissance satellite to date."[44]

There could be little doubt that, at least in one area, the new version of Gambit was superior to its predecessor. While targets read out did not compare favorably with best Gambit-1 results to that time (targets had been photographed on the 27th flight of Gambit-1), the improvement which could ultimately be expected was indicated by progress recorded since Gambit-1's first flight.[45]

Gambit-3 program plans dated from December 1963 and January 1964. The initial launch of the system had then been scheduled for 1 July 1966. Almost a full year later, booster changes caused a schedule revision to reflect a new first-launch date of 28 July 1966. In the event, launch was postponed by 24 hours. In sum, a schedule established 134 weeks beforehand proved to be only four weeks in error. In retrospect, that did not seem an enormously significant achievement—and was not. But if considered together with the budgetary record of Gambit-3, it represented a unique achievement in those years of major program slippage, budget overruns, and performance shortfalls throughout the Department of Defense. No DoD program of the 1960's that approached Gambit-3 in terms of technical advance, gross cost, or scheduling stringency achieved such a record. The caution that Gambit-3 was really no more than an advanced version of a system already in operation had no real relevance. Even though its mission was the same as that of Gambit-1, Gambit-3 provided substantial advances in resolution and orbital life. It required development of a new camera with more than twice the focal length of the original, new film, batteries, fuel plumbing—and ultimately a new booster. The performance objectives of 1 resolution and ten-day orbital life constituted goals beyond reasonable expectations of earlier years—and when Gambit-3 itself matured, even those goals were surpassed.**

The fate of the Gambit-1 system was, of course, markedly influenced by the success of early Gambit-3 flights. By June 1966, shortly before the first Gambit-3 flight, there were clear indications that the first lot of Gambit-3 systems probably would not produce much better resolution than the (now) well regarded Gambit-1. United States Intelligence Board (USIB) approval of Hexagon development some two months earlier (April 1966) further complicated the problem; early expectations of Hexagon availability, though unreasonably optimistic, made it seem advisable to begin near term conversion of the Gambit-1 launch pad for use by the Hexagon. The certainty that Gambit-1 would be entirely replaced by Gambit-3 made it difficult to keep the Gambit-1 development-manufacturing team adequately motivated—and, indeed, made it unlikely that a Gambit-1 capability could be maintained at all, given the increasing needs of the Gambit-3 program. If Gambit-3 were even moderately successful, the need for completing the planned purchase of 16 additional Gambit-1 systems would vanish; if, on the other hand, Gambit-3 encountered early operational problems, keeping a reserve Gambit-1 capability in being would be essential to continuance of the surveillance mission. Dr. Flax was extremely reluctant to cancel any planned Gambit-1 launchings until Gambit-3 had actually demonstrated a "reasonable" level of capability.[46] Despite Dr. Flax's reluctance, the Director of Central Intelligence, Richard Helms, felt that the combined total of 20 Gambit (counting both Gambit-1 and Gambit-3) in fiscal year 1967 was too much. He pointed out that the schedule developed for fiscal 1967 dated from January 1966. At that time, Gambit's had experienced a succession of catastrophic failures—but there had been no Gambit-1 failures during 1966. Further, Gambit-1 had recorded remarkable advances in orbital life and coverage during that year. Finally, the first Gambit-3 had flown successfully and the considerations which underlay the January decision were quite obsolete. On 17 August, therefore, the Executive Committee for the National Reconnaissance Program decided to delete four of the scheduled Gambit-1 flights.[47]

USIB's Committee on Overhead Reconnaissance in September 1966 proposed that nine Gambit-1 and eight Gambit-3 missions be conducted during fiscal year 1967. USIB, as a whole, somewhat reluctantly accepted the recommendation of its subcommittee, several members expressing concern that success in all of the scheduled missions would cause the exploitation elements of the intelligence community to be swamped in high resolution photographs. Many of those USIB members also favored contined use of Gambit-1 rather than Gambit-3 because of the apparently greater cost of the newer system.

For the moment, the decision to proceed with a mix of Gambit-1 and Gambit-3 systems during the 12 months starting in July 1966 was permitted to stand unchanged. The convincing argument, advanced by the staff of the National Reconnaissance Office, was that the better roll maneuverability and longer orbital lifetime of the Gambit-3 system in combination with a resolution that, at worst, would be at least as good as that of Gambit-1, were sufficient justifications for proceeding rapidly to total reliance on Gambit-3.

---

* Resolution was not obtained with the Gambit-1 system until its last two flights, numbers 37 and 38, in May and June 1967.

** The 22nd flight achieved resolution of and an orbital life of 27 days was later recorded.

One of the growing national concerns during the period of Gambit-3 development was quick reaction to crises. As early as January 1965 Dr. McMillan had informed Secretary McNamara of impending studies of a quick reaction capability for Gambit-3.[48] While funds had been allocated in the fiscal 1966 budget for this purpose, little work had been done—the money was held back until actual results of Gambit-3 operations could be weighed.

The tangible proof which had been lacking during fiscal year 1966 became available shortly after the first Gambit-3 flight, at the beginning of fiscal year 1967. On 17 August 1966, the NRP Executive Committee approved in principle the modification of Gambit-3 by inclusion of multiple recovery vehicles and extended orbital life. The costs involved were two years in the future and the technology involved was, by that time, low risk. Long life and multiple reentry capsules would satisfy the need.[49]

The premise of long-life, multiple capsule systems was that urgently required photographs would be taken and returned for analysis while more routine surveillance duties were still performed, exposed film being fed into a second recovery vehicle. But another requirement for quick reaction was that either a satellite be on-orbit when the need arose or that one could be launched quickly. One was largely a matter of luck and the other was, during 1966, ultimately constrained by the rate of production of reconnaissance satellites.

While the first G-3 flight had been largely meant to prove the capability of the camera and gain some experience with orbital maneuvers, the second flight was intended to demonstrate some extension of the satellite's orbital life and to test, as exhaustively as possible, the roll joint and the backup systems.* The camera system was still the pacing item in scheduling, and on-pad testing had to be compressed slightly when the acceptance testing slipped a week. The slippage affected the Astro-Position Terrain Camera, which was not a critical item for initial flights of Gambit-3. When that subsystem failed its acceptance tests (a looper malfunction due to a broken roller bracket), it was disabled and the second Gambit-3 system was launched without its services on 28 September 1966.[50]

The vehicle achieved a correct orbit and began its planned seven day primary mission, programmed to photograph almost targets—half again as many as the first flight.[51] A tape recorder malfunction was the only disturbance of the primary mission. It failed during orbit 82, but recovered thereafter. The capsule was recovered, uneventfully, on orbit 115.[52]

The roll joint was exercised extensively during the primary mission (943 cycles), and in solo flight (546 cycles). Only one malfunction occurred (during solo) and it was corrected via backup systems.[53]

The importance of the backup systems being well understood by the project office after Gambit-1 experience, a total of 1552 backup system operations were completed during the solo portion of the mission.

The results of the second Gambit-3 flight were mixed. While the mechanical operation of the satellite was impeccable, the camera system did not score as well. Photography for the second flight was poorer than that of the first, a variance in the quality of the optics being blamed.[54] The outcome was puzzling, however, because preflight tests had indicated that the optics for the second Gambit-3 were slightly superior to those of the first. Best ground resolution was about 36 inches. Nevertheless, the second Gambit-3 success reinforced the growing conviction of NRP managers that too many Gambit's were in the flight schedule.** Shortly after recovery of the film from the second Gambit-3 flight, the photographic payload for the third flight began acceptance tests toward a 27 October delivery in anticipation of a 1 November launch date. But the payload experience of the second flight induced an extension of testing for the third, causing the flight to be slipped until mid-December. The extended test procedure was adopted as the norm for all future payloads.[55]

The vehicle was readied during the second week in December and launched on the 14th. The orbit achieved was similar to the two previous flights, although somewhat more eccentric.[56] Two pre-planned orbit adjustments during the primary flight lowered the perigee from 82.6 nautical miles to approximately 76 miles.[57] During the flight, spurious commands generated through one channel of the extended Command System troubled flight controllers, but actually affected operations only during three orbits (28 to 31), during which most of the operational photographic take was lost. Thereafter, changed operational procedures overcame the

---

* In September 1966, Colonel King left the Gambit office for a new assignment. In some respects, the timing was unfortunate; King had been the prime mover in Gambit development for several years, and the second Gambit-3 mission still was pending. General Martin, therefore, arranged for King's continued availability to oversee the second and third Gambit-3 missions—a task that had precedence over any he incurred in his new assignment.

** Two issues underlay the overlapping flight schedule; cost and continuity of coverage. The success of both Gambit flight programs during 1966 created pressures to reduce the total number of flights. Contemporary schedules called for one Gambit-1 per month in the period October 1966 to June 1967 plus Gambit-3's in November 1966, January 1967, and one per month thereafter. Three of the planned Gambit-1 flights were eliminated. Because of the optics problem, the Gambit-3 flight planned for November was slipped to December, and the January, March and May 1967 flights of Gambit-3 eliminated.

difficulty. The Astro-Position Terrain Camera which had malfunctioned on flight one and been deliberately disabled on the second flight, again malfunctioned, experiencing intermittent operation of the shutter mechanism.

The mission returned "best resolution" on a new ultra-thin-base film (still not the high-speed film earlier promised) that extended the available film load from roughly 3000 feet to about 5000 feet. That increased film capacity largely accounted for an improvement in the numbers of targets programmed and read out. Another contributing factor was mission length: eight days.[58] Gambit-1 had yet to achieve that goal. But Gambit-3 had still not been extended to its full capabilities. The project directors were carefully adhering to the "proceed slowly" rule for development flights.[59]

The objectives of development flights included proof of hardware capability as well as learning how to extract the maximum return from the hardware. The fourth flight of Gambit-3 principally served the latter end. The flight was scheduled for 21 February 1967. No major problems were encountered during acceptance testing, but preflight checkout disclosed an out-of-specification condition in the inertial reference package of the guidance section that ultimately caused a three-day slip in the launch.[60]

By shortly after noon of 24 February 1967, the satellite control facility at Vandenberg sent the cryptic "Nominal ascent" message to the dozen or so stations waiting for word.

The fourth mission of Gambit-3 thereafter proceeded with few on-orbit problems. Flight controllers confidently predicted that the photographic take would be as good as that of the first Gambit-3 flight, thus far not duplicated.[61]

Some anomalies did occur, of course. Erratic behavior of the focus sensor, incorrect timing of the backup roll joint drives during the solo flight, and erratic spacing between photographic frames were the most notable.

Double images and run-through of unexposed film resulting from frame spacing faults absorbed the greatest amount of administrative and technical energy during the days following recovery of the film.[62]

Unhappily, best resolution was on the order of 27 inches, a development that—in light of good preflight test results and promising on-orbit performances—demanded explanation. The only obvious difficulty experienced in the camera section acceptance test had involved determination of the best point of focus.

"Best focus" was established by adjusting the platen with reference to a tri-bar target similar to those photographed from orbit. Skilled technicians focused the optics repetitively until a large body of data on the "best electrical focus" distance had been built up. The data were then plotted. All previous experience, both with Gambit-1 and Gambit-3 had been that the resulting plot formed a unimodal distribution. The mode, or peak, was chosen as the point of best focus.

Because of an aberration in the optics for the fourth Gambit-3, the distribution of points of best focus had two peaks. The better of the two also displayed double imagery. In an attempt to use the higher resolution focal distance designers improved the lens, which eliminated the secondary image. The optics were flown in this configuration despite a supplementary test using a point target which indicated that the lower resolution peak was unambiguously correct. On-orbit, defocus experiments, unhappily, agreed with the results of the point-target procedure.

But the wrong "best point of focus" had been selected before launch. As a result of that experience, all future optical assemblies in which an aberration had been detected were focused by use of a point target rather than the more common tri-bar target. A cause for chagrin was a set of calculations which showed that resolution of the mission photography would have been as good as that obtained on the first mission had the correct focal plane been chosen.[63]

In other respects the mission had been successful. Despite the double imagery and comparatively low resolution, the intelligence value of the photography was considered to be high. Exposure was excellent for more than 90 percent of the exposed frames. Free for the first time of malfunctions, the terrain camera registered ground resolutions between 100 and 120 feet, more than adequate for mapping purposes. The stellar camera element again experienced film fogging, although not as badly as had been the case on the first three missions.[64]

By the time hardware acceptance had begun for the fifth flight of Gambit-3 there was reason to anticipate a very successful mission. Acceptance testing of the cameras ran more smoothly than ever before.[65] Procedures for determining the best plane of focus seemed quite adequate and tests showed the camera to have a well defined stable point of best focus with a resolution potential of 60 lines per millimeter.[66] While nominal lens resolution was inferior to that of the fourth flight article, it seemed more than adequate. The optics tested better than those of the first Gambit-3 camera, that which had achieved the best resolution of the series. General Martin reported to his Washington counterpart, General Berg that, "The recommendation is a very strong 'Let's go!'"[67] No significant problems had appeared in acceptance testing, delivery, or mating of the components on the launch pad. A slip of one day from the planned 25 April 1967 launch date apparently was caused by weather conditions in the launch area.*

Lift-off occurred at exactly 10:00 a.m. Pacific Standard Time on 26 April 1967. Before the message could be sent to that effect, however, the second stage booster, Agena, and camera section had impacted in the Pacific Ocean southeast of Hawaii. The first notice of failure mentioned a possible second stage Titan failure adding, "Injection into orbit is questionable at this time." The final message from Point Arguello that day spoke for itself: "Water impact due to low thrust of second stage of Titan booster. Negative acquisition at downrange ship on ascent. Further report will be issued only if additional information becomes available." That day, at least, none did.

Such information as eventually became available indicated that thrust chamber pressure had dropped by 40 percent shortly after second-stage ignition. Maximum velocity was 8000 feet-per-second, too slow for injection. Suspecting that the failure was due to a fuel line blockage, Martin instituted more stringent inspection procedures on all tanks, lines and pumps. No more could be done.[68]

The sixth and final Gambit-3 development flight had originally been planned for 6 June 1967. Following the failure of mission five, it was rescheduled for 20 June. Acceptance testing and prelaunch checkout proceeded smoothly, no major anomalies being encountered.

Early in June the NRO staff in Washington recommended consideration of a new flight pattern for Gambit-3 with the object of exploiting the higher sun angles characteristic of summer months. It was accepted and made effective for the sixth Gambit-3 mission.

In the past, Gambit cameras had normally been operated only on north-to-south orbital passes. By launching in the early morning, instead of early afternoon, photographic operations could take place on the ascending portion of the orbit as well. In consequence, some targets, those which fell in the ascending track, could be photographed twice a day instead of only once. Making such an adjustment also required that the latitude of perigee be shifted northward so that the satellite would be closer to the earth during the ascending portion of its orbit. Although resolution would be acceptable on both sides of the orbit trace, targets in southern latitudes would register poorer resolution than usual because of the camera's higher altitude when it flew over them. The tradeoff was acceptable because of better coverage of "more interesting" Soviet and Chinese target sites during the ascending portion of the orbit.[69]

Detecting a slight anomaly in the optics, technicians set the "best electrical focus" one mil away from the calculated best plane of focus. Modal distribution seemed slightly abnormal, and because the platen position was subject to slight shocks during ascent, a one-mil, out-of-focus condition of the platen seemed the best compromise between expected resolution and a potential out-of focus condition.**

A change in checkout procedures for the sixth Gambit-3 provided yet another increase in total film capacity. Final checks of the system before a flight normally consumed from 200 to 300 feet of film. To offset that loss, Eastman overloaded the spool; the first 300 feet of film extended above the spool flange. The flange kept film from slipping loose during launch and injection, but it was not necessary in the controlled ground experiments used during checkout. (The first two Gambit-3 flights had carried only 2600 feet of film, as did most of the later Gambit-1 vehicles. The introduction of ultra-thin base film had allowed an increase to 4700 feet. With the adoption of the new checkout procedure, Gambit-3 could be launched with almost 5000 feet of unexpended film.)

The mission began on schedule, early in the morning of 20 June 1967. All ascent events proceeded normally until 60 seconds before Titan second stage engine cutoff. At that moment, part of the ablative skirt of the engine blew off. The resulting aerodynamic asymmetry slowed the vehicle's acceleration, and injection velocity was 88 feet per second lower than programmed. Apogee was more than 46 nautical miles lower than planned, and perigee was about two miles low. Mission controllers called on the Agena propulsion system to correct the orbit, but during that maneuver (on orbit 32) the Agena rocket engine suffered a chamber pressure loss. Fortunately, it did not affect the ultimate success of the mission.[70]

---

* The principal documents which cover launches are Quarterly Program Reviews of Special Projects and the routine memo from the Director, SAFSP, to Director, NRO (Mission Summary). Both of these documents were concerned with the most important feature of the mission—its failure—to the exclusion of all other details. However, high winds in the launch area were common during March and April, accounting for similar one-day slips on other launches.

** The conservatism of the choice was warranted by flight experience. Analysis revealed that the electrical focus had shifted a half mil from its intended position.

Indeed, notwithstanding such mishaps, the flight proceeded so well that controllers decided to extend orbital life from the planned eight days to ten. An Agena reburn during orbit 98 provided the essential repositioning impulse.

A malfunctioning relay brought on roll joint operating peculiarities during and after the 64th orbit. Flight controllers spent some 14 hours in identifying the source of the problem before attempting a correction. The first fix attempted was to use the crab servo to work around incorrect roll-joint movements. After three orbits during which smear was inevitable, new programming instructions went aloft. Not until orbit 112 was the problem wholly overcome.[71]

The photographic products of the sixth Gambit-3 mission were well worth the effort required to make the flight successful. Two extra days of flight and the extra film contributed to a doubling of the number of targets photographed and read out. Resolution showed less variability than in earlier Gambit-3 operations, as did exposure. Photography was considered generally to be the best achieved by Gambit-3 to that time.[72]

Analysis of the film results by the National Photographic Interpretation Center indicated that the decision to adopt a new orbit had been wise. NRO headquarters reported: "All evaluations received to date regarding the intelligence effectiveness and overall scale and exposure quality of NSN 4306 have been highly complementary... it is requested that the capability for high northern latitude perigee placement and early morning launch be retained and requirements be examined for each Msn during the high solar northern declination months for application of this type orbit. We wish to commend your development of the 4306 type orbit and its contribution to increased intelligence return in terms of operational flexibility and cost effectiveness."[73]

Dr. Flax was somewhat less euphoric. Reporting to the Deputy Secretary of Defense, Flax noted that, "In general, system performance was excellent (after six flights) except for the optical sensor which is, of course, the most critical element of the system." Resolution fell well short of the planned resolution. But in all other respects, Gambit-3 had to be considered a striking success. As compared to Gambit-1, the newer system had achieved its development goals—excepting resolution—in about one-third as much mission time.[74]

Results obtained during the first year of Gambit-3 operations fully justified the decision to rely on that system. Although the newer high-resolution film was not yet available, Gambit-3 produced a "best resolution" of (as compared to the "best resolution" of Gambit-1), an average resolution that was slightly better, and a substantially greater rate of film exposure. (Owing to various limitations on maneuvering during the first three Gambit-3 missions, in deference to research and development test objectives, Gambit-3 averaged coverage of only slightly more targets than Gambit-1, but on a "best missions" basis the Gambit-3 system covered half again as many critical targets.)

After the second Gambit-3 flight had appropriately demonstrated the general technical capability of that system, Dr. Flax cancelled the final five Gambit-1 missions and diverted the boosters to other assignments.[75] There was abundant evidence that the interpretation facilities would be overloaded were the full complement of Gambit-1 and Gambit-3 vehicles to be flown in what remained of the year. After the sixth Gambit-3 mission it was clear to all concerned that the system would produce better returns than Gambit-1 could ever hope for and that it was wasteful of resources to continue production of Gambit-1. On 30 June 1967, Flax cancelled the Gambit-1 program. At the time, seven additional Gambit-1 systems were on contract, but only two were approaching completion (work on the other five having been earlier curtailed). Even though the improved film on which Gambit-3 expectations were partly based still had not been perfected, Gambit-3 recorded a "best resolution" during its ninth flight (mission 4534), in October 1967.

Results of the adoption of Gambit-3 were evident in more generally acclaimed ways than "best resolution" alone. The U.S. knew, for instance, almost precisely at what rate the new Soviet T-62 tank was being delivered to Soviet tank regiments stationed along the Chinese border, and similar findings were reported for a surprisingly wide variety of aircraft, missiles, and ships. That the newer Gambit returned pictures that made it possible for U.S. interpreters to assess was a sufficient comment on the improvement that had resulted from its introduction.[76]

Policy, Administration and Further Development

The sixth flight marked the end of the development flight program for Gambit-3. In terms of mission operations, however, that milestone was mostly distinguishable by a slight decrease in experimentation during flight. The first four flights of Gambit-3 had been clearly developmental in that they had limited objectives. Malfunctions in some non-critical components were tolerated, for instance, as had also been true of Gambit-I in a comparable phase. A significant difference in the experience of the two programs was the amount of attention devoted to the transition from developmental status to full operational readiness. For Gambit-1 the pace of that transition was an issue of considerable concern generated partly by political considerations (the mounting CIA-NRO differences of the time) but also by the complete lack of any alternative means of obtaining high resolution overflight photography. The NRO probably could not have survived another program failure, so Greer and McMillan were exquisitely circumspect in their deliberate progress toward full operational readiness. But precisely because all previous programs had been failures, the intelligence community was enormously impatient to get early operational returns from Gambit. Largely because of Greer's stubborn insistence on step-by-step progress, Gambit-1 had an unprecedented run of early successes. Major problems when they occurred, came toward the mid-point of Gambit-1's operational life, at a time when intelligence specialists had become accustomed to a steady output of high-quality photography and had come to assume, almost placidly, that any Gambit problems would be solved quickly and without disrupting the flow of information to photo interpreters. Gambit-1 experienced mid-term problems because of manufacturing and test process faults; Greer and King had successfully insured against the survival of any disabling design defects. In Gambit-3 the design verification process proved out in the earlier program was supplemented by the manufacturing process emphasis that General Martin had applied to Gambit-I. Defects in Gambit-3 tended to be random, the product of oversight or accident rather than any failures of process.

Development of the camera systems of both Gambits was a never ending process. Development-during-production expenditures for Gambit-1 camera systems (identified as such by their being entered in the books as non-recurring costs) far exceeded initial development expenditures. That also became true, for Gambit-3. That a potential for continual improvement existed testified to the excellence of the basic design of the two camera systems. (The same had been true of Corona, though in a somewhat different fashion.)

During the period covered by the development flights of Gambit-3, the NRO was working its way toward an optimal mix of reconnaissance vehicles. During the early days of Gambit-1, the entire photo-satellite reconnaissance effort was dependent on two systems, Gambit and Corona, each based on technology of the late 1950's. Gambit-3 represented a vehicle at least as reliable as the second generation Corona, and like that system promptly became an apt subject for modification and performance improvement. Among the developments which concerned Gambit-3 were readout technology, extended operational life, and multiple film return capability. Readout and multiple reentry implied longer system orbital life and expanded coverage of ground targets. Imagery readout had been a goal of satellite reconnaissance for two decades—originally because no obvious alternative ways of retrieving orbital photography were available. The introduction of recoverable film capsules, with Corona, and the indifferent quality of returns from early readout systems (Samos E-1 and E-2) had relegated readout to a research enterprise by 1962, however. During 1966, the concept of readout was again raised to system design status, and in one proposed application Gambit-3 became the vehicle. Various technical considerations made it only marginally attractive at the time, however, and in November 1967 the NRP Executive Committee decided against any near-term application to Gambit-3. It recurred more and more frequently to consideration of new system possibilities thereafter, however.[77]

Although readout was rejected for the time, long life and multiple reentry capability had become approved Gambit-3 goals earlier in the year. Neither involved the technological and financial risk of imagery readout. Fortunately, owing mostly to McMillan, the Titan booster of Gambit-3 had life capability the early Gambit-3 systems did not need. The addition of a second reentry vehicle, more film capacity, and the attendant modifications to film transport, cutting and other mechanisms would considerably increase Gambit-3's weight. Except for "crisis reconnaissance" operations, a capability never required though often proposed, a second recovery vehicle would be redundant if the total time on orbit for the camera were no greater than with one capsule. Extended orbit of life would require more batteries and greater quantities of control gas and rocket fuel. Additional lift capability through the addition of small strap-on solid rocket engines to the Titan III was an option which had been foreseen early in the Gambit-3 program (and provided for in construction of the launching pad at Point Arguello). Formal approval for the development of long-life, multiple-reentry vehicle capability for Gambit-3 was obtained in August 1966.[78] The change would add almost to Gambit-3 costs in fiscal 1968.[79]

By November 1967, the question of schedule overlap between Gambit-1 and Gambit-3 was conditioned by the continuing success of Gambit-3 operations. For a time late in the year, the CIA suggested that Gambit-3 rather than Gambit-1 schedules should be curtailed. Absolute cost comparisons were all in favor of Gambit-1—if resolution were disregarded.[80]

The success of the first two Gambit-3's made it apparent to the United States Intelligence Board that with three satellite systems routinely returning large quantities of photographic intelligence, the interpretation task was becoming, for the first time, a constraint on operations.

Photo interpreters were not alone in finding it difficult to deal with three successful satellites at the same time. The Satellite Operation Center (SOC) in the Pentagon (Washington's principal interface with STC, Vandenberg) was also feeling the pinch. Major General James T. Stewart, Director of the NRO staff, complained that, "The SOC is barely able to cope with Gambit and/or Gambit Cubed today in a manual operation." Addition of Hexagon to their burden would swamp them, as would any appreciable improvement in Corona capability.[81]

By early December 1967, USIB was considering a reduction in coverage requirements as a way out of the difficulty. In combination with extended life for Gambit-3, that implied steadily decreasing launch rates for the future. At the launch rate then prevailing, the 16 Gambit-3 systems then on order would last through April 1970. At the usage rate suggested by USIB, only 12 would be required. After allowing for backup vehicles to replace possible failures, the NRP Executive Committee decided to buy 14 rather than 16 new Gambit-3's.[82] (In the event, 16 were actually used by June 1969, after which the dual-reentry vehicle [double bucket] version of Gambit-3 became standard.)

The recurrent Gambit-3 launch-rate question was nominally resolved in January 1967;* six launches were scheduled for fiscal 1967, ten for fiscal 1968, nine for fiscal 1969, and seven each year thereafter. That decision reflected a cutback in scheduled launches of seven vehicles through fiscal 1969 and two per year thereafter. Uncertainty concerning the provision of standby and backup vehicles was countered by a decision to consider reserve vehicles as hedges against expanded intelligence requirements rather than as insurance against the catastrophic failure of a mission. Thus readiness within two months became the standard for reserve vehicles, rather than readiness within two to three weeks (as would have been the case if protection against mission failure consequences were intended).[83]

In a separate effort to conserve funds, the NRO altered its earlier disaster recovery policy for Point Arguello. Against the possibility of a disabling launch pad disaster, the NRO had planned to modify a second pad to back up the primary Gambit-3 launch pad (PALC II, Pad 3). That commitment was revoked in February 1967 in favor of a policy of quick rebuilding and repair in the event of launch stand damage. The newer approach had the undeniable advantage of being less costly, particularly if no severe damage occurred. It invoked a degree of greater risk, of course, but in that no occasion for either major repair or the use of an alternate launch stand arose through the end of 1972, the policy subsequently justified itself.[84]

The only major shortcoming that inhibited Gambit-3 by late 1967 was the inadequacy of the camera system. Despite the fact that it was better than that of Gambit-1 it was not yet sufficiently better to justify Gambit-3 development—and could not be so rated until the original requirement for resolution was actually satisfied. That achievement depended on real progress in developing high-resolution, high-speed film, improved substrate materials, better manufacturing process, and improvements in the final preparation of the optical surfaces.

All of the cameras flown in the first six flights carried optics whose substrate material was fused silica. All available alternative materials had been rejected by the beginning of 1967. It was less expensive and easier to manufacture than fused silica substrates. Those were important considerations; the 1967 rate of production was so low that it prevented the accumulation of enough spares to permit diversion of mirrors to the development of faster polishing techniques—a circularity which would require considerable effort to break.

During the first quarter of 1967, six promising but undersized blanks were sent to Perkin Elmer and Eastman for finishing. Unfortunately, the castings were marred by opacity and structural defects eliminated those faults in the next batch of blanks, but the castings still were overweight, and polishing techniques were not adequate to the need.[85] Still, the casting process seemed to have progressed past the period of experimentation. Late in July, Colonel

---
* It was reopened annually, however.

Lew Allen left a note on General Berg's desk* which provided a good indication of progress: "After many disappointments it appears that we have a good 72" blank. The press operated flawlessly... We still need to hold a reservation until full clean-up and interior inspection is done, but our experience leads us to feel success is here. Hooray!"[86]

Notwithstanding such progress, the resolution specified for Gambit-3 still had not been realized. The urgency of achieving better than resolution in Gambit-3 operations had its own rationale in system requirements, but it was reinforced by activity elsewhere, particularly in Hexagon programs. In some respects, it seemed to be a prospective replacement for Gambit-3.

The elusive goal of photo-satellite reconnaissance was a good resolution of small objects. It was a threshold number for photo interpreters, who foresaw in it marvelous possibilities for obtaining hitherto unobtainable information. By 1967, some seasoned veterans of the National Reconnaissance Program had begun to suggest that Gambit-3 might never achieve even the basic resolution for which it had been designed, but that it would actually return imagery of better resolution. Commenting on a draft study by Colonel Lee Battle, former Corona program manager, General Berg conceded that, "[basic resolution] ... may be just too tough a development problem. But compare that to the theoretical statement that 'we will get better at the outset' from it. I want to see better resolution so badly, I become completely vexed with anyone talking it down. But I also realize that getting it depends a heck of a lot on the G^3 learning curve and experience..."[87]

Uneasiness about Gambit-3's prospects also surfaced during an August 1967 meeting of the Reconnaissance Panel of the President's Scientific Advisory Committee (PSAC). Although the nominal topic was the program, the PSAC had also scheduled a one-hour briefing on the status of Gambit-3 optics. In the event, it lasted three hours, counting the extended discussion it generated. Dr. Richard L. Garwin sharply criticized several aspects of the Gambit-3 program: program management, lack of technical expertise at Eastman Kodak, and the faltering progress in improving the resolution provided by the first Gambit-3 mission to the better resolution specified. General Berg cautioned Dr. Flax that the PSAC might well pursue the topic at greater length in another session, but the threat never materialized. While members of the PSAC were fully aware of Gambit-3's difficulties, and concerned about them, most did not accept Garwin's pessimistic views. The NRO staff concluded that Garwin's criticisms were in part based on incomplete information.[88]

Notwithstanding the failure of Gambit-3 to satisfy early expectations for resolution, its demonstrated operational capability was sufficiently satisfying by the sixth flight to warrant final cancellation of Gambit-1. In consequence, several camera systems and attendant flight equipment was freed for distribution to other programs. Most went to NASA to support the Surveyor and Lunar Orbiter photography programs, although two were crated and stored for display at "some future time" when security permitted.

Making Gambit-1 cameras available to NASA created its security problems. In order to conceal the potential of the now outmoded Gambit-1, enormously better than anything in NASA's inventory, the NRO specified that the equipment be called the "Lunar Mapping and Survey System" (LMSS) and be used only for lunar photography—and that products be presented to the public in a way that would make it impossible to determine resolution. That could be accomplished—in theory—by failing to disclose the altitude from which photographs were taken, thus concealing both scale and definition. In that fashion, the need for invoking Talent-Keyhole security procedures would be obviated.[89]

As it turned out, the LMSS camera was never used for lunar photography. But the ownership of equipment so advanced presented new opportunities, and new temptations, to NASA. Some NASA people began to argue for the use of the Gambit-1 camera in the Earth Resources Survey and Earth Sensing programs. The product would be photography with at least two-foot resolution in a program that was closely monitored by a large community of scientists who were unwitting of the achievements of covert reconnaissance. The National Security Council had earlier ruled that NASA could not release photography with better than 60-foot resolution. There seemed no feasible way of concealing Gambit-1 capability if the system were used at any scientifically reasonable orbital altitude—although NASA was not willing to concede that point. With the vigorous support of Dr. Donald Hornig, the President's Science Advisor, the NRO responded to NASA's enthusiasm by stony refusal; the issue was allowed to die of its own weight.[90]

While such events were proceeding, and while development flights of Gambit-3 still were continuing, designers had begun to work seriously on the development of a dual-recovery-vehicle version of the system. Film capacity was the limiting factor in Gambit-3's orbital life, the ten-day film supply generally

---

* Allen, later head of the NRO staff, and still later head of the Special Projects Directorate, was then in charge of technical development in the NRO staff; Brigadier General R. A. Berg headed the NRO staff at the time.

being used up long before the system was otherwise exhausted. On the seventh mission, for instance, even after completion of a solo mode operation and deboost of the Agena control section, remaining battery life was sufficient for five additional days of life, and enough control gas remained for another 28 days on orbit.[91]

The small quantity of film available, relative to other expendables, evoked considerable ingenuity among flight planners intent on securing maximum returns from the single-bucket version of Gambit-3. That ingenuity was rewarded by significant and steady increases in target coverage from the first to the last (22nd) single-bucket Gambit-3.

The development period allowed for the double-bucket Gambit was fundamentally determined by the rate at which the original Gambit-3 systems were used up. Following the purchase of six development systems, the NRO bought 16 additional single-bucket Gambit-3's for a total of 22. The last was expected to be launched before April 1969, which thus became the scheduled initial launch date for the first double-bucket Gambit-3. But small schedule slippages in development of the double-bucket Gambit could be, and were, offset by stretch outs in single-bucket launch schedules. In late October 1967, less than a year after work had begun, it became evident that July 1969 rather than April 1969 was the probable first flight date for the initial double-bucket Gambit. Launches of the remaining single-bucket satellites were stretched out to cover the gap.[92] (The relatively late finding that a double-bucket Gambit would require a more comprehensive data handling and command subsystem had an appreciable influence on the schedule slip)[93]

While the events peripheral to the development and operational quality of Gambit-3 continued to intrude on the program, and in many instances tended to distract attention from the central purpose of that program, the technical features, intelligence potential, and on-orbit performance of the system were themselves in a constant state of change. Such change was neither random in nature nor wholly responsive to problems encountered in operation; both technological progress and mission performance were somewhat unpredictable influences on program status.

Several improvements in the Gambit-3 system became feasible almost concurrently with the shift to a dual-recovery subsystem. The major innovation was a new optical system.[94] The new system provided greater focal length, a flatter field, and improved color correction. Both resolution quality and the variability of resolution would improve thereby.[95]

The introduction of improved film was also a constant of the Gambit-3 program. A medium-speed (E.I 3.6), thin-base film had been used during the first two flights. The same emulsion, on a thinner base, was carried on flights three through thirteen. By the time of the 14th flight, the faster emulsion (E.I 6.0) originally scheduled for the first Gambit-3 flight was deemed ready for use. Because it was somewhat less satisfactory than had been anticipated, the slower emulsion was reintroduced, temporarily, for flights 20 through 27, after which an improved version of the higher-speed emulsion on ultra-thin base was adopted. It exceeded original expectations.

Two other major advances in Gambit-3 technology originated in research performed for the program. The first was a solution to the stubbornly intractable problem of obtaining a fully satisfactory optical substrate material. Half a dozen materials had been tried and rejected for one reason or another during the course of Gambit-3 development. The only materials that still seemed to offer promise by the time it approached readiness were fused silica, used on virtually all flights of the single-bucket series. Both materials were candidates for the substrate. Because fused silica had always been an interim solution to the Gambit-3 mirror substrate problem, Gambit lens designers seized eagerly on the new material. In a sense, it was the last real hope for mirror improvement; none of the feasible alternatives had the desirable qualities of materials. Not until flight 42 of Gambit-3 was the new substrate used.

The second innovation was the Titan III "long tank" modification. The 32nd flight of Gambit-3, marking the introduction of system, was also the point of adoption of "long tank" Titan III boosters. (The excess lift capacity of the Titan III-X had been sufficient to permit an uncomplicated conversion from single- to double-bucket versions.)

In early 1968, USIB again suggested changes in coverage requirements and Gambit-3 launch frequencies, explicitly proposing a reduction to seven flights during fiscal year 1969.[96] The NRO had earlier concluded that eight vehicles would have to be procured to insure seven successes (and nine to insure eight successes). Eastman's optics production capacity was still limited, and eight systems were a more realistic prospect than nine. Secondly, by that time only eight more single-bucket vehicles remained on contract and the availability date for the first double-bucket Gambit-3 was somewhat uncertain. Stretching out the single-bucket Gambit-3 launch schedule was the least painful way of protecting against a gap in coverage.[97]

In the event, the first flight of the double-bucket Gambit-3 did not occur until 23 August 1969, almost five months after it was originally scheduled and two months after the last single bucket Gambit flew. Thus a short hiatus in intelligence return did occur.[98]

FLIGHT PROGRAM – VEHICLES 7 TO 22

Acceptance testing for the flight equipment to be used on the seventh Gambit-3 mission went smoothly. The payload was accepted with a total of nine performance waivers, of which only one, a shortfall in resolution, was important. It was also recurrent and persistent, continuing to be "the major performance waiver."[99]

During preflight testing on the launch pad a faulty command in the test command tape caused a failure in the roll joint primary motor electronics. The entire component had to be replaced, causing a slip of one day in the launch. The vehicle finally lifted off shortly after nine in the morning of 16 August 1967. The early launch time was, as in the case of the sixth flight, used in order to take advantage of high sun angles still available during August.

Despite the "nominal launch" message from Vandenberg, the injection was not nominal.[100] The initial perigee was six nautical miles high (80.3 miles) due to an injection velocity 25 feet per second higher than planned. The satellite was allowed to remain in high orbit until its 31st revolution, when it had decayed to a perigee of 79.8 nautical miles. At that point an orbit adjustment brought it some five miles closer to earth.

Even before the non-nominal orbit had been detected, ascent telemetry indicated that the roll joint separator had not functioned, causing it to be "locked up" and unable to traverse from side to side along the flight path. Backup systems successfully freed the roll joint almost four hours after launch. Minor problems also occurred with the battery thermostat in the recovery vehicle making it necessary for flight controllers to turn off the battery heater power for most of the mission. Some channels of the Extended Command System also became troublesome, but neither failure degraded the mission. Two further orbital adjusts were carried out on orbits 81 and 162.[101]

The number of targets programmed and read out were both extremely high. That so many targets actually were photographed was partly due to favorable cloud conditions over areas of interest. Ground resolution as measured from photographs came out nearly as good as the first Gambit-3 flight.[102]

There were no delays in acceptance or checkout of flight equipment for the eighth flight of Gambit-3, and that pattern persisted through the mission. Mission eight was one of three in the first group of 22 for which no major on-orbit anomalies were recorded. Of several minor anomalies, however, two were significant. As with mission 4307, the "nominal launch" signal sent immediately after launch on 19 September 1967 was a premature indicator of mission normality, injection error causing the perigee to be almost four miles lower than predicted. Because it was within acceptable tolerances, however, no orbital adjustment was performed until perigee had decayed to less than 71 nautical miles, on the third day of the mission. Another orbit adjustment proved necessary three days later. An eccentric film wrap prevented primary camera operation for several minutes during the first pass over the Soviet Union, but was quickly corrected. (It was one of a very few camera malfunctions in the early operation of Gambit-3.)[103]

Because of the season, mission 4308 was launched later in the day than its two predecessors, there being no possibility of taking advantage of high sun angles. Instead, the satellite trace was altered for better coverage of targets at lower latitudes (descending) than had been obtained in recent missions. Target coverage suffered from the change, but an offset was the achievement of resolution, equal to the best yet obtained. The average quality of the photography was superior to that of any earlier Gambit mission.[104]

During acceptance testing of the photographic payload section for the ninth Gambit-3 mission, tests indicated that resolution would be better than with earlier optics. However, intermittent operation of the platen drive motor caused a two-week slip in delivery to allow for retrofit. The eventual effect was a two-week slippage of the launch itself.[105]

An additional one-day delay in launch occurred with the detection of a propellant leak in the Titan second stage. Launch finally occurred on 25 October 1967.[106]

Launch trajectory and orbital injection were near perfect. The Astro-Position Terrain Camera experienced a minor malfunction and the last half day of photography was lost due to the failure of the film take-up system. (Some 200 feet of film remained in the camera when the recovery vehicle separated for reentry.) Otherwise, troubles were few.

Upon analysis, the photography proved to be the best ever recovered from orbit. Measured resolution was and less and eight percent of the frames were incorrectly exposed.[107] CIA Director, Richard

Helms, was moved to compliment Gambit's recent photography, characterizing it as a rich source of "extremely important intelligence."[108]

Pressures for target selection changes reached the Gambit Project Staff from various groups within the intelligence community. *(Paragraph and Associated Endnote (109) Redacted.)*

Flax's reaction was to schedule a compilation of coverage experience so to "enable us to relate planned aircraft and drone coverage and, if necessary, to plan special or additional G missions, assuming the requirements were 'hard.'" The resulting study indicated that satisfaction of nominal USIB requirements for it would require no fewer than 20 Gambit missions a year. Since, at the time, USIB was in the process of trying to decide whether Gambit missions should total six or seven annually. Flax's interest in distinguishing between "requirements" and "hard requirements" was warranted.[110]

Actually, communications between the NRO and the CIA in that matter dealt with mapping and charting applications of Gambit. Apparently not all in the intelligence community appreciated that although Gambit-1 often had unexpended film at mission's end (mission event selection procedures for that system were relatively unsophisticated in terms of later procedures), Gambit-3 relied on far better programming routines and "wasted" only a very small amount of primary film, and absolutely no indexing film whatever.[111]

The tenth Gambit flight was little distinguished from the ninth. Resolution achieved was the same, although overall film quality was somewhat lower. The Extended Command System failed during its test cycle at Vandenberg, but was quickly replaced by the unit intended for the next flight. Launch occurred on 5 December 1967. Ascent and injection appeared to be nominal; although the orbit was actually slightly high. As a result, an orbit adjust programmed for the 80th orbit was rescheduled for orbit 33 and was supplemented by another adjust on orbit 96. The primary reason for orbital adjustments was target optimization, however, and not recovery from an anomalous orbit. Small, non-critical malfunctions occurred during the mission in the Extended Command System and the terrain camera part of the APTC.[112]

Early in 1968, one of the few hardware changes during the single-bucket flight program took effect—completion of the development of a redundant attitude control system (RAGS) as a backup for the primary system used in the satellite control section. While no failure of the primary system had occurred in Gambit, thus far, the project officers were wary of such an occurrence. Experience with the Gambit-1 vehicle had been considerably different, and since the weight of a redundant system could be easily accommodated in Gambit-3, it seemed wise to make the change.[113]

Gambit-3's eleventh mission was not a happy one. The vehicle seemed plagued with problems from its beginnings. Acceptance testing for the photographic payload section began on 8 December 1967 but was halted six days later when a telemetry unit malfunctioned and had to be replaced. The replacement unit failed on 2 January 1968. A third unit was installed and the payload was accepted, with waivers, on 4 January. (The major waiver was again for the optics, which were still considerably below specification. Initial resolution tests were so bad for the unit that the optical axes of the Ross Corrector and the primary mirrors were realigned using interferometry. The realignment yielded a significant improvement in resolution so the system was accepted at a double-pass resolution.

The satellite control section had comparable problems. During on-pad checkout at Vandenberg, a line surge through one of the breakout boxes resulted in burned wires in the separation controller. It took two days—passed on as a launch slippage— to remove and replace the unit and recheck the remainder of the vehicle.[114]

Finally, on 18 January 1968, mission number 11 was begun.[115] Launch and injection were almost nominal again. But the orbit was again low and had to be corrected during the 32nd orbit. Another orbit adjust was performed on orbit 96. Much earlier, the primary viewport doors had refused to open. Fortunately, the backup system worked properly.

On the tenth day of flight, recovery procedures were put in train, but although telemetry signals were mostly positive, planes in the recovery area never made contact. Nor did tracking stations. The recovery system had functioned correctly up to and including dispatch of the coded message indicating that the parachute cover had been ejected, but two minutes later all telemetry ended. It could only be assumed that the capsule had reentered the atmosphere without benefit of parachute. A search was begun without much hope of finding the capsule. It ended some thirty hours later with the assumption that reentry velocity had been sufficiently high to destroy the capsule on impact with the water.

Despite such mishaps, the mission was not a complete loss. After separation of the capsule, the Agena satellite control section continued in orbit for an additional seven days, logging thereby the longest time on orbit yet experienced for the system. The purpose of the extended solo mission was to demonstrate the capability of the control system to operate over the total time required in support of the double-bucket Gambit. The control section was deboosted after a total of 17 operational days on orbit.[116]

Another matter conditioned by the expectation of longer life beginning with vehicle number 23 was the capacity of the roll joint, the link between the power supply stored in the Agena and the payload section. The Agena, stabilized by its own attitude control system, received commands from tracking and guidance stations and translated these into roll maneuvers that would allow the camera to point at targets to the side of its trace. Because "interesting" targets fell on the satellite's trace only by accident (i.e., rarely), virtually every target to be photographed was the cause of a roll maneuver.

The ultimate reason for increasing Gambit-3's time on orbit was to obtain an increased number of good photographs of important targets. A fully successful mission could require many separate roll maneuvers. The roll joint used in the first eleven flights had been capable of performing only rolls. A new drive system installed for the 12th vehicle had a capacity of rolls. It represented the first step in moving toward the capability for the double-bucket Gambit.[117]

The only other major equipment change on Gambit-3 number 12 was a newly designed parachute. Its installation caused a launch date slippage of seven days. Otherwise, optics were predominant matters of concern, although the electronics for the Astro-Position Terrain Camera had to be removed and reworked before the payload section was acceptable. Primary optics were again the subject of a performance waiver, a procedure almost ritual for Gambit-3 by that time. Although continuing progress in polishing techniques and improvements in the use of interferometry had steadily improved system resolution, the Gambit-3 system as a whole was not yet capable of satisfying original specifications. Tacit acknowledgement of that reality was signaled by the decision to disable the short range compensation mechanism after it operated erratically during tests. The trouble was traced to a drive motor. A failure during flight could cause platen adjustment loss and mission failure, so the compensator was disabled. The penalties were inconsequential; the optics were still so imperfect that correctly making the fine adjustments for which the slant range compensation device was intended would have no detectable effect on the quality of mission photography.

All was ready by 12 March 1968 but, as had happened before, high winds at Point Arguello forced a further slip of one day. On 13 March, fifteen minutes after the launch window opened, the mission began.

Launch and injection were again nominal.[118] The only significant flight problem was failure of the terrain-camera element of the Astro-Position Terrain Camera on orbit number four. But because the stellar camera operated throughout the mission, indexing for the primary camera was adequate.

Analysis of mission telemetry provided the lead to the solution of a major problem of camera operations. The difference between predicted and actual resolution for previous missions had been much greater than expected.* The point of best focus determined in preflight tests seemed always to be off by an unacceptable margin. By the 12th flight, a considerable amount of experience had been built up through secondary flight objective experiments with the focus adjustment. On the 12th flight it was noticed that the focus wandered during the course of each revolution between +.001 inches at the start of the orbit and -.001 inches at the end.** Analysis established that heating of the front surface of the stereo mirror by the earth albedo was responsible. Flight controllers attempted to stabilize the mirror by cooling it for ten minutes during the dark-side portion of each orbit. The experiment was only partly successful, but it led to a far better understanding of thermal influences on focus adjustments.[119] Ultimately, it would become possible to minimize heat-induced focus shift by regulating the time and degree of camera door and viewport opening during various phases of the mission.

After ten days of flight, the capsule was separated—and the new parachute worked. The recovery vehicle was air recovered on 23 March 1968. The general level of photography was better than on any previous mission.[120]

A more serious special coverage requirement inquiry was dealt with on mission number 13. In response to a request made by the Director, Joint Reconnaissance Center, the program office prepared a description of coverage from April 1967 to March 1968, corresponding to Gambit-3 missions five to twelve. During that period, targets that had been programmed had been either fully covered, partially covered, or had poor coverage. Some targets were programmed for coverage during fiscal 1969. Designing a mission for enhanced coverage could increase the programmed coverage of targets while still satisfying USIB requirements for the rest of the world on that mission.[121]

---

* See Analysis of Gambit (110) Project.

** That is, the distance between lens and platen shifted in this fashion, causing the focus to change.

In the end, that pattern was imposed on the forthcoming 13th Gambit-3 mission. ("Requirement is max access to the main target plus unique access to at least 60 percent of the remaining targets.")[122] The change caused no major delay in the acceptance process, although it was necessary to inhibit the activity of the platen adjust motor in order to avoid possible catastrophic failure resulting from malfunction of this component. The vehicle was ready on 16 April 1968 but high winds in the launch area again caused a delay of one day, to 17 April.

The mission was "nominal" from start to finish; there were no significant malfunctions on orbit. Because of the stringent coverage requirements, and the consequent need to fly a near cyclic orbit, two orbital adjusts were carried out (orbits 64 and 113). Two more took place during the subsequent two-day solo mission (to study the thermal effects of low altitude flight, an experiment aimed at demonstrating capabilities of the control section in anticipation of more demanding requirements for the operation of the double-bucket Gambit).[123]

The results of the mission were encouraging. Resolution again improved. In addition, more targets were both programmed and read out than ever before. The coverage was the subject of a congratulatory message from Washington: "Please accept the congratulations of Dr. Flax and the NRO staff for achieving the highest readout of intelligence targets in the Gambit Program to date."[124]

If anything, the 14th mission of Gambit-3 was more smoothly run than the 13th. Acceptance testing and checkout cycles were uneventful. Launch occurred on time on 5 June 1968, early in the morning (as had the previous flight). The mission flew for its planned ten days and recovery was also nominal. The only failure on orbit involved a tape recorder intended to record data on malfunctions elsewhere in the vehicle—but there were none to record.[125]

Of greatest interest were the results of the mission. The resolution achieved was again photographic quality being slightly poorer than had been registered on mission 13. The number of intelligence targets read out was greater, however, even though fewer had been programmed. (More targets were cloud free.)[126]

Experiments conducted during the 14th flight of Gambit-3 had the additional benefit of demonstrating that target coverage could be enlarged without altering flight parameters. Reducing the length of each "burst" of photography would reduce the quantity of film used in photographing areas surrounding the real target. As a product of improvements in target location data (generated largely by the Corona program) and corollary improvements in Gambit position data, the need for "insurance" footage had gradually diminished. By the time of the 14th flight of Gambit-3, enough was known about true position on orbit to achieve a far greater degree of control over the camera than had been possible at the beginning of the program. That knowledge was applied during the flight in experimental reduction of burst times. The results of the experiment were good "...this is a successful test and could be applied in future missions with the exclusion of the highest priority targets..."[127]

Mission 15 of Gambit-3 used short burst times as standard. The result was another dramatic increase in the number of targets programmed and read out, despite slightly heavier cloud cover.

The planned launch date had been 23 July 1968, but problems in checkout of the payload caused a slip to 6 August. The mission was near nominal with no major malfunctions. A quick-look estimate of resolution was later revised upward, equivalent to experience in the previous mission.[128]

Seven more Gambit-3 vehicles remained before the first of double-bucket vehicles appeared. They were flown between September 1968 and June 1969. System changes and experiments carried out during that time were mostly aimed at proving capabilities needed in the double-bucket version of Gambit-3.

The first new item of equipment, and possibly the most significant addition to Gambit-3 during the period, was a redundant attitude control system (RAGS). The first was installed in the 16th vehicle and was the subject of tests during the solo portion of that flight. Timing was exquisite: a failure of the primary stabilization system on flight 17 forced reliance on the still experimental redundant system, after which the mission proceeded to a successful conclusion. "As a matter of fact," General Berg told Dr. Flax, "John (General Martin) tells me the vehicle stability is better. In this case we can chalk one up for the home team!"[129]

Flight number 16 also included some experimentation with color photography as well as the introduction of a third version of the roll joint, a redundant system to be used in the double-bucket Gambit-3. The color photography was quite successful, achieving both good color correction and good resolution.[130] Such an achievement was noteworthy; the earlier Gambit-1 system had been designed to produce resolution of two to three feet using the best available monochromatic film. Here was more evidence of the influence of changing technology on Gambit.

Flights 18 and 19 were marked by extremely high energy orbits which caused a short mission in the first instance and degraded photography in the second. *(Paragraph and Associated Endnotes (131 and 132) Redacted.)*

During the last three flights of the original Gambit-3, resolution improved during mission 20, mission 21, and then mission 22. The resolution specified for Gambit-3 had finally been surpassed by the last camera to be flown in that series. The final two flights of the single-bucket Gambit also set records in target coverage with both programmed and readout targets during the 22nd flight.

Photographic quality benefited from several innovations introduced during the flight program of the Gambit-3 vehicle. They included the introduction of ultra-thin-base film over "over-flange wrap" (for preflight tests), ascending and descending photography during summer months, and reduction in burst times made possible by more accurate position data on the vehicle and its targets. Resolution was also improved through the continual development of better optical surfaces, more accurate polishing techniques and closer tolerance testing. Other contributing factors included improved understanding of the influence of thermal variation on the optical surfaces during flight (and the ability to control these through operation of the viewport doors), and better control of camera dwell times and stereo mirror flip time. One area where improvement was lacking, where some real deterioration was noticeable, was the effect of cloud cover. The Gambit project staff did not see it as their responsibility to alter the weather, but they were charged with avoiding poor weather to the extent that other factors permitted. Perhaps the 417 weather satellite had provided all the assistance in its capability; perhaps Gambit-3 was unlucky in that respect. But in any case, target obscuration by clouds continued as a major inhibitor of complete mission success.

The Gambit-3 flight of 3 June 1969, mission number 22, had been skipped to compensate for delayed availability of the first double-bucket vehicle. In the event, the mission schedule adjustment was not sufficient, and last-minute problems caused a gap in coverage of almost two months between the 22nd and 23rd flights of Gambit-3. In general, the results of the 23rd mission and following flights of the improved Gambit-3 system proved to be well worth the extra wait.[133]

FLIGHT HISTORY OF GAMBIT-3 BLOCK II (DOUBLE-BUCKET)

The first group of double-bucket Gambit-3's (referred to here as the Block II vehicles) experienced three kinds of capability improvement. Resolution increased from a previous best and the improvement in average resolution was even more dramatic. Operational longevity increased from ten days to 27 days. In the entire period there were only three catastrophic failures: the second recovery vehicle was lost on flights 25 and 27, and mission 35 was a total loss. It was not that technology took a sudden spurt. Rather, a succession of modifications conceived and tested over several years coincided in their effect. The Gambit-3 system matured appreciably between August 1969 and September 1972.[134]

The first group of Block II flights included missions numbered 4323 through 4327. Each was planned as a 14-day mission, a week of photographic operations being allocated to each of the two recovery vehicles (SRV's). Of the five flights, the third and the fifth were marked by catastrophic failures; a parachute failed to open for the second recovery vehicle of mission 4325, and an Extended Command System failure on mission 4327 prevented recovery of the second SRV from that satellite. Although these vehicles brought back 50 to 100 percent more photography than the last of the single bucket Gambit-3's, their best resolution was not better. Average resolution improved slightly.

Other than the introduction of the second SRV, only two major hardware changes distinguished the first lot of Block II Gambit-3 vehicles: a battery was added to extend the orbital life of the vehicle, and an improved reserve attitude control system was introduced.

No major malfunctions marked the first Block II flight. An excessively high orbit (408 nautical miles instead of 220) marred the second (4324). Because of a relay malfunction, the vehicle control assembly failed to shut down the main engine during ascent. Happily, the failure did not degrade mission success, orbital correction being possible through use of the secondary propulsion system rather than the primary gas supply. Remote tracking stations had some difficulty in picking up the vehicle's transponder frequency, which caused some delay in loading commands and a loss of some photography. But the effect was minor. The post-flight discovery of a dome nut tucked into the folds of the recovered film prompted redesign of those nuts for future flights, but again there was no mission effect.

The recovery parachute on the second capsule of mission 4325 failed and the capsule sank before recovery personnel could reach it. Flight 4326 encountered no major problems, but registered the worst resolution seen for a year and a half.

Mission 4327 was routine through recovery of the first capsule, although the air-catch crew was somewhat startled to discover that the ablative shield had failed to separate upon parachute deployment. Still both the parachute and the air-catch harness were able to sustain the extra weight, so no harm resulted. But shortly after film began to pan to the second capsule, the heater for the clock in the extended command system began malfunctioning. In the absence of that component, the propulsion systems of the vehicle could no longer be precisely controlled and little could be done to govern the impact point of the second capsule when it deboosted. Program controllers decided to attempt an emergency recovery, but it was unsuccessful and the second capsule, carrying one-day's film was lost.

The first 18-day flight of Gambit-3 was scheduled for 18 August 1970, one year after the first double-bucket Gambit-3 mission. No major hardware changes had been necessary to extend the 14-day life of the earlier vehicles to 18 days. Mission 4328 remained operational only for 16 days, however, it was called down early. Loose thermal tape interfered with operation of the horizon sensor and detracted from photographic operations in a minor way, but the total take was significantly greater than on any previous mission.

In the next three flights, between 23 October 1970 and 11 May 1971, there were no hardware changes and no malfunctions of any consequence. Mission 4330, begun on 21 January 1971, achieved, for only the second time among the Block II vehicles, a resolution as good as the best of the earlier Gambit-3's. It was also the first of the Block II vehicles to undergo an atmospheric survivability test (VAST). Despite close tracking, no debris could be located on this or any of the subsequent three tests, conducted at the conclusion of missions 4331, 4332, and 4334, reassuring intelligence specialists that no revealing bits and pieces of Gambit hardware could survive reentry and thus provide clues to the system's composition or capability.[135] Later events were to show that confidence to be wholly unwarranted, but that was for the future.

The last of the second lot of Block II Gambit-3 operations, began on 22 April 1971 and continued for 19 days. Vehicle 4331 achieved a best resolution and exposed more film than any earlier Gambit.

Mission 4332, begun on 12 August 1971, two years after the first Block II launch, marked a new watershed in the maturation of the Gambit program. The orbital life of 4332 was extended from 18 to 22 days by the introduction of a tenth one-kilovolt battery and other minor system modifications. Far more significant, however, was the first operational use of equipment, under development for several years. The system improved best resolution from its earlier peak to a higher percent performance improvement.

Several hardware changes in the four flights which followed led to still longer orbital life, improved stability, increased maneuver capacity, and still better photography. The mission 4333 Gambit-3 used, for the first time, a new type of battery that enabled it to stay on orbit for an extra two days, bringing total mission life to 24 days. The booster and second stage had reached their lift capacity with the introduction of the newer—and heavier—batteries, however. In order to carry still more batteries, therefore, it was necessary to increase the thrust of the Agena upper stage by the use of High Density Acid (HDA) in place of the standard oxidizer. That modification allowed either an increase in orbital payload of 125 to 150 pounds, or an increase in the vehicle's inclination capability. On vehicle 4334, which included this modification, a fifth "magnum" battery was used. The satellite was launched on 17 March 1972.

Mission 4335 included more major hardware changes than any other Block II vehicle. A Gemini Uniform Mixture Ratio (GUMR) engine was installed in the secondary propulsion system which further enhanced lift capacity by an additional 200 pounds, or alternatively, allowed spacecraft inclination to approach eight degrees. The introduction of a latching solenoid valve in the Backup Stabilization System (BUSS), allowed the use of surplus BUSS gas in the primary control system, which effectively increased the number of rolls that could be performed. Finally, some of the extra lift capacity was utilized to completely replace all of the original batteries with the newer, heavier, "magnum" models, bringing the total of the more powerful batteries to ten and increasing orbital capability life to 30 days. Unfortunately, a defective pneumatic regulator prevented the delivery of control gas during the ascent stage of the launch of mission 4335. In consequence, the satellite failed to orbit and the entire mission was a failure.

While mission 4335 did not include a VAST experiment, an attempt was made, as on all flights, to identify the impact point for whatever debris might remain after reentry. The predicted impact point for 4335 was somewhere over South Africa. Ground tracking stations lost contact with the disabled spacecraft during its descent, however, so no search was attempted. Considering the few instances in which debris had survived reentry, that seemed safe enough.

Almost five months after mission 4335 had been launched, a Gambit-cleared employee of the Aerospace Corporation visiting his firm's London offices, heard from a coworker, some "interesting space material" had been recovered by the British earlier that year. The employee was interested, so the coworker arranged to revisit the Royal Aircraft Establishment at Farnborough, where he had first seen the debris on a laboratory bench.

The employee found three objects on open view, a spherical titanium pressure vessel about a foot in diameter, some circuit boards of U.S. manufacture, and several chunks of glass which, together, formed a pie-shaped wedge with a ten-inch edge. The glass was backed by a characteristic of Gambit optics. All the pieces had been found on farmland within a five-mile radius some 75 miles north of London. Eyewitness accounts indicated that they had come down about 20 May 1972.

Convinced that he had viewed debris from a Gambit vehicle, he privately alerted the project office in California. The resulting flurry of telephone calls and memoranda led to informal arrangements for recovery of the residue, the transfer being arranged through friendly RAF contacts. Subsequent analysis confirmed that the debris had indeed survived the breakup and reentry of Gambit 4335[136] —which cast some doubt on the findings of Project VAST and further lessened confidence in impact-point predictions.*

The offending pneumatic regulator was replaced for mission 4336 and later Gambit-3 flights, but no other hardware changes were made.[137] Gambit 4336, the last Block II system, was launched on 1 September 1972 and remained in orbit for 27 days of photography, after which the satellite control section was exercised for an additional day. All of the new hardware introduced in the previous mission functioned correctly, and none of the several minor malfunctions had any substantial impact on mission performance.

Three distinct advantages arose from the addition of a second satellite recovery vehicle to Gambit-3. The increase in photographic coverage was the most obvious—although, in fact, photographic days on orbit did not dramatically increase merely by the addition. Perhaps more important, the second capsule enhanced the quick reaction capability of the Gambit; if need be, the first film batch could be recovered as soon as vital photographs had been obtained without forcing an end to the entire mission. Finally, a "Quick Look" team assembled to perform rapid analysis of the output of the first lot of recovered film could direct focus or ephemeris corrections that enhanced the quality of the second lot of film.

Gambit-3 vehicles 23 through 36, comprising the "Block II" buy, incorporated continuing changes in hardware through their period of use. Refinement of operational procedures had been the most important source of improvement for "Block I" vehicles, although modifications, subsystem changes, and optical system development also occurred. But hardware changes during "Block II" flights were chiefly responsible for orbital life extension from an original 14 days to an eventual 27 days. For Block II, procedures changed mostly as hardware changed. Major design changes included the introduction of a long tank Titan booster, introduction of a modified fuel system for the Agena, detail changes which allowed improved utilization of orbital control gas, and the introduction of larger and more numerous batteries. In order to take advantage of the extended life thus provided, flight controllers had to place greater reliance on short-burst times and on early morning launches. The reduction of burst times, film pads, and frame sizes required a number of improvements in hardware and software. Data collected throughout the Gambit-3 program were used to improve target location and orbital parameter accuracies. Reduced burst times thus became possible. The introduction of the MOD IV Command Subsystem permitted more precise calculation and control of ephemeris changes.[138] The reduction in burst times was so effective that over the initial period of Gambit-3 operations, the average frame length decreased from 2.68 feet in 1966 to 1.29 feet in 1969, the first year of Block II operations. By the end of 1972, average frame length had dropped to slightly more than six inches.[139]

Such accommodations initially extracted a price in degraded resolution. Ascending photography required higher average altitudes, and the brevity of burst times was limited by requirements for stabilization after a roll maneuver (settling time), and flip time for the stereo mirror. Smeared photographs resulted if bursts were too brief and too frequent.[140] To some degree, these disparities were increasingly offset by continued improvements in film. Indeed, film quality so outpaced the mechanical capabilities of the vehicle, principally the roll joint, that by the time Block III changes were planned (vehicles after 36), a principal addition was a new roll joint capable of supporting maneuvers during the mission.[141]

---

* Five years earlier, an entire Corona capsule had survived random reentry and landed, largely intact, in Venezuela. See Volume I for details.

The quality of Gambit-3 photography was also degraded in early flights by high frequency banding and focus shifts, both which were, at first, poorly understood. Banding, troublesome for the first lot of Gambit-3 satellites, was eventually attributed to a lack of smoothness in the film drive. The problem was corrected by stiffening the drive shaft between the motor and the camera platen. Focus shift, had also been identified earlier, but not until mission 4332 was it demonstrated conclusively that focus shifts occurred because of temperature changes brought on by camera door settings. Setting primary camera doors at appropriate angles to provide shade helped to regulate the temperature gradients across the lens.[142]

Perhaps the most important Gambit-3 improvement during the Block II program was the introduction of new equipment. Old equipment had difficulties with the original optical system of Gambit-3. The first nine double-bucket vehicles had resolution slightly better than the last few single-bucket Gambits, and resolution actually degraded on occasion. Early Block II Gambit-3 vehicles used the same cameras as Block I systems, but various problems associated with Block II design changes caused resolution to worsen for several flights. Average resolution had only returned to the level achieved late in the Gambit-3 Block I flights on mission 32. The new equipment permitted resolution to surpass the previous best, achieved only three times in the entire Gambit program.[143]

With the launch of the first double-bucket Gambit-3 on 23 August 1969, the mission life of that system doubled and the cost per target covered dropped significantly, though not proportionately. By December 1972, fourteen of the double-bucket Gambit systems had been launched; of the 28 potential film recovery opportunities thus created, 24 reached fruition. On one occasion the recovery parachute failed to deploy (a mishap that also marked the first flight of Hexagon in mid-1971, when two of four recovery parachutes deployed improperly and one payload was lost). Later in 1970 a command system failure sent a recovery vehicle back into the atmosphere still attached to the orbital vehicle, and in May 1972 an Agena malfunction ended in failure to achieve orbit. Following somewhat after the introduction of the twin-capsule version of Gambit, the new equipment was incorporated and "best resolution" immediately went from a normal range to better. Target coverage increased. None of the high-resolution films routinely used by 1972, had existed in 1969; film improvement was a major factor in Gambit-3's extended scope of coverage and vastly improved resolution.[144]

The first Block III Gambit-3, 4337, was launched on 21 December 1972. Recovery of the second SRV on 2.2 January made 4337 the longest Gambit mission yet flown (31 days). Best resolution was not quite matching the achievement of mission 34, but average resolution was as good as any earlier recorded. *(Paragraph and Associated Endnote (145) Redacted.)*

The next mission of Gambit-3, operation number 4338, was plagued by malfunctions which inhibited operations to some extent, although they did not cause substantial failure. Problems began before launch with the discovery of an unreliable power supply for the camera's electronic focusing apparatus. The launch date slipped by two days in consequence, to 16 May 1973.

Launch itself was uneventful, but problems with the Astro-Position Terrain Camera (APTC) developed almost immediately. The terrain camera portion produced no usable frames before the 232nd revolution. Owing to extreme underexposure, intermittent failure of the entire APTC occurred after the 300th revolution. Finally, not all the photography which had been planned could be accomplished; only 129 feet of film could be recovered because the take-up reel in the second recovery vehicle had overfilled.[146]

(Paragraph and Associated Endnotes (147 and 148) Redacted.)

McLucas telephoned Dr. J. R. Schlessinger, then Director of Central Intelligence (and thus chairman of the Executive Committee for the National Reconnaissance Program), at 9:20 that morning. Unable to contact Schlessinger directly, McLucas left a message: unless advised to the contrary by 11:00, McLucas proposed to allow the recovery of the first capsule to proceed as Bradburn had directed—that afternoon. The cost, he advised, would be about 15 percent of programmed coverage of relatively low priority targets and inability to use some 1500 feet of film.

No objections appeared. At 4:01 (Eastern time) on the afternoon of 21 May, a C-130 circling east of Hawaii caught the capsule in its descent and headed for its base. Twelve hours later, after a nonstop flight from Hawaii, the spooled film reached Eastman's Rochester, New York, laboratories. Despooling, developing, and locating and inspecting the critical frames required seven hours, during which the atmosphere shifted from hope to gloom to elation.

Although the first phase of mission 4338 lost nine days on orbit, the second was flown for seven days longer than had been planned. It might have been flown for an additional two days except for the expenditure of control gas in adjusting Gambit-3's orbit. Ultimately, all but a small portion of the film flown on 4338 was exposed and all but 25 feet loaded into recovery vehicles. No high-priority targets and few low priority targets programmed for coverage before launch were missed. Advancing the launch date for the next Gambit-3 on schedule would offset any coverage gap.

The last mission of fiscal year 1973, mission 4339, was begun on 26 June 1973. It proved to be a disappointing anticlimax to the high achievement of 4338. Some 12 seconds after the early morning launch from Vandenberg Air Force Base, the main fuel tank of the Titan ruptured. The debris fell into the Pacific Ocean south of Vandenberg.

With the launch of the first double-bucket Gambit-3 on 23 August 1969, the useful single-mission intelligence return virtually doubled. The single-bucket version had carried 5000 feet of film, the double-bucket Gambit-3 carried almost 10, 000 feet. The number of roll maneuvers the system could perform increased to 7000 from an initial 1250. Time on orbit increased from ten to fourteen days. Principally due to two operational changes brought about during the first 2.2 flights of Gambit-3, the number of targets acquired was enormously increased. Those changes included early morning launch during summer months (to enable both ascending and descending photography), and the use of shorter burst times (i.e., less film per target).[149]

By 1973, the Gambit program had moved into its fourth generation—Gambit-1, Gambit-3, the double-bucket Gambit-3, and the "Block III" Gambit-3 (the 37th Gambit-3). That newest satellite was initially capable of several camera operations, each one accompanied by a roll maneuver. It could resolve images on a side of several nautical miles and from heights of several nautical miles—a feasible operational altitude.

The achievements of the Gambit program from its inception in 1963 to 1973 were varied, significant, and in many cases, dramatic. One that was often overlooked was cost. Although Gambit photography improved in resolution over those years, the photographs themselves became less expensive by several orders of magnitude. *(Paragraph and Associated Endnote (150) Redacted.)*

Endnotes

1. Rpt, SAFSP Quarterly Program Review, 10 July 1964; (hereafter cited as QPR, with date).

2. Memo, E. M. Purcell, Chm, Reconnaissance Panel, to DCI, Jul 63, subj: Panel for Future Satellite Reconnaissance Operations; memo, B. McMillan, DNRO, to Dir CIA, 11 Sep 63, subj: Implementation of Purcell Panel Recommendations, both in SAFSS files.

3. MFR, MGen R. E. Greer, Dir/SP, 15 Aug 63, subj: Plans for Ultra-High Resolution Satellite Reconnaissance.

4. Rpt, "Preliminary Development Plan for Advanced Gambit System (Program G)," prep by SAFSP, Vol I, 4 Feb 64.

5. Memo, MGen R. E. Greer, Dir/SP to Col W. G. King, Dir/Gambit Ofc, 13 Dec 63; subj: G3.

6. Msg, MGen R. E. Greer, Dir/SP to BGen J. L. Martin, Dir/NRO Staff, 27 Dec 63.

7. Memo, MGen R. E. Greer, Dir/SP to Col W. G. King, Dir/Gambit Ofc, 2 Jan 64, subj: G 3 , msg, BGen J. L. Martin, Dir/NRO Staff, to Greer, 3 Jan 64.

8. Ltr, MGen R. E. Greer, Dir/SP, to Col W. G. King, Dir/ Gambit Ofc, 7 Jan 64, subj: Letter of Instructions; msg, B. McMillan, DNRO, to Greer, 8 Jan 64; ltr, Greer to King, 8 Jan 64, subj: Appointment of a Special Evaluation Board G 3 , in SP files.

9. Prelim Dev Plan for Advanced Gambit Sys (Program G3), Vol I, 4 Feb 64.

10. Ibid, pp 14-1, 14-2.

11. Msg, BGen J. L. Martin, Dir/NRO staff, to MGen R. E. Greer, Dir/SP, 10 Feb 64; msg, Greer to Dr. B. McMillan, DNRO, 24 Feb 64; ltr, EKC to Greer, 28 Feb 64, subj: Proposal for Recoverable Satellite Reconnaissance System G3.

12. QPR, 10 Jun 64.

13. Memo, MGen R. E. Greer, Dir/SP, to Col W. G. King, Dir/Gambit Ofc, 28 May 64, subj: G3.

14. Msg, BGen J. L. Martin, Dir/NRO Staff, MGen R. E. Greer, Dir SP, 2 Jan 64; msg, Greer to Dr. B. McMillan, DNRO, 3 Jan 64; msg, Martin to Greer, 4 Jan 64, in SP files.

15. Ltr, MGen R. E. Greer, Dir/SP, to LMSC, 5 Jun 64; ltr, Greer to GE, 5 Jun 64; memo Greer to Col W. G. King, et al, 9 Jun 68, subj:G 3 SCS Parallel Program; msg, BGen J. L. Martin to MGen R. E. Greer, 1 Jul 64.

16. QPR, 30 Sep 64.

17. QPR, 31 Dec 64; memo BGen W. G. King, Dir/SP, to Dr J. L. McLucas, DNRO, 28 Apr 70, subj: Analysis of Gambit (110) Project.

18. Msg, MGen R. E. Greer, Dir/SP, to Dr. B. McMillan, DNRO, 22 Oct 64; msg, McMillan to Greer, 30 Dec 64.

19. Memo, B. McMillan, DNRO, to DepSecDef, 4 Jan 65, subj: Milestones for Gambit-3, in DNRO files.

20. Msg, 100, , BGen J. T. Stewart, Dir/NRO Staff, to MGen R. E. Greer, Dir/SP, 8 Jan 65; msg, SAFSP to EKC, 21 Jan 65.

21. QPR, 31 Mar 65.

22. Msg, BGen J. T. Stewart, Dir/NRO Staff, to AFSC, 4 Feb 65.

23. Msg, SAFSS to EKC, 4 Mar 65.

24. QPR, 31 Mar 65.

25. Msg, MGen R. E. Greer, Dir /SP, to Dr. B. McMillan, DNRO, 8 Mar 65; msg, , BGen J. T. Stewart, Dir/NRO Staff to Greer, 9 Mar 65.

26. QPR, 31 Mar 65.

27. QPR, 30 Jun 65. There is some ambiguity in this source, however. In the overview section, compression of test schedules is cited as the reason the EKC slippage did not impact on the initial launch: in the technical status section, EKC is described as pacing the entire G-3 Program.

28. QPR, 30 Jun 65.

29. *(Text Reference was Redacted) QPR, 30 Sep 65.*

30. QPR, 30 Jun 65.

31. *(Text Reference was Redacted) Ibid.*

32. QPR, 30 Sep 65; 31 Dec 65.

33. QPR, 30 Sep 65.

34. QPR, 31 Dec 65.

35. See procurement resume in memo, King to McLucas, 28 Apr 70, attachment 4. The exception was the camera; 22 flight articles were purchased at one time.

36. QPR, 31 Dec 65.

37. Rpt, "A Review of the Gambit-3 Program," Feb 68, in NRO staff files.

38. QPR, 31 Mar 66.

39. Rpt, "A Review of the Gambit-3 Program," Feb 68.

40. *(Text Reference and Endnote was Redacted)*

41. Msgs, 29 Jul 66, 29 Jul 66, 2 Aug 66, all Sat Ops Center to Dir, SP. Note that "z" time is Greenwich Mean Time (GMT). When daylight savings time is in effect, there are seven hours difference between GMT and Pacific Time; when not, the difference is eight hours.

42. EKC, "Addendum I to the Final Flight Evaluation Report for Flight No. 1, Operation No. 3014, on 29 July-6 August 1966, 3 Oct 66." These reports referred to hereafter as "EKC Evaluation Flight No____."

43. QPR, 30 Sep 66; memo, King to McLucas, 28 Apr 70.

44. EKC Evaluation Flight No. 1, p 5.

45. Memo, Col C. T. Smith, Gambit Ofc, to BGen J. L. Martin, Dir/SP, 24 Aug 67, subj: Analysis of Gambit Project.

46. Msg, Dr. A. H. Flax, DNRO to BGen J. L. Martin, Dir/SP, 10 Aug 66.

47. Min, NRO ExCom Mtg, 17 Aug 66.

48. Memo, Dr. B. McMillan, DNRO to R. McNamara, SecDef, 11 Jan 65, subj: Quick Reaction Surveillance System.

49. Min on NRP ExCom Mtg, 17 Aug 66.

50. Msg, 28 Sep 66; QPR, 30 Sep 66.

51. Msg, 28 Sep 66.

52. Memo, King to McLucas, 28 Apr 70, Atch 1, Tbl 1, and Atch 3.

53. Msg, Tower 1434, 31 Aug 66.

54. QPR, 31 Dec 66.

55. QPR, 30 Sep 66, 31 Dec 66; memo, BGen J. L. Martin, Dir/SP, to Dr. A. H. Flax, DNRO, 2 Feb 67, subj: GAMBIT Cubed Mission Summary. Hereafter, these memoranda will be cited as "Mission Summary, Flight No._____."

56. Msg, 14 Dec 66.

57. Msg, 23 Dec 66.

58. Memo, King to McLucas, 28 Apr 70, Atch 1, Tbl 1, and main report.

59. See Ch XIII.

60. Mission Summary No. 4, 25 Apr 67.

61. Msg, 24 Feb 67; msg, 5 Mar 67.

62. Msg, 25 Apr 67; msg, 23 Mar 67; Mission Summary No. 4, 23 Mar 67.

63. Msg, 23 Mar 67; Mission Summary No. 4, 25 Apr 67; QPR, 31 Mar 67; Addendum I to EKC Evaluation, Flight No. 4, 11 May 67. See also MFR, BGen R. A. Berg, Dir/NRO Staff, 23 Mar 67, subj: Telephone conversation with Gen Martin.

64. EKC Evaluation, Flight No. 4, 11 May 67, p 3-1.

65. QPR, 30 Jun 67. "The contractor appears to have significantly improved his component and manufacturing quality control," was the project office evaluation.

66. QPR, 31 Mar 67.

67. MFR, Berg, 23 Mar 67.

68. Msg, 26 Apr 67; Mission Summary, No. 5; QPR, 30 Jun 67; msgs, 11 Apr 67 (Mission parameters), 26 Apr 67 (count), 26 Apr 67 (launch and first failure indication), 26 Apr 67.

69. Mission Summary No. 6, 27 Jul 67; msg, SAFSS to STC, 1 Jun 67; interview, 16 Apr 73.

70. QPR, 30 Jim 67; Mission Summary No. 6, 22 Jul 67.

71. Msg, 3 Jul 67.

72. These data were provided by of Special Projects. In addition, the Mission Summaries cited here are all from files maintained by the only ones extant in the case of early G-3 missions. In addition to the targets mentioned above, additional ones were photographed outside the programmed areas. See msg, 31 Jul 67.

73. Msg, 25 Jul 67.

74. Memo, Dr. A. H. Flax, DNRO, to DepSecDef, 6 Jul 67, subj: National Reconnaissance Program (NRP) Issues and Pending Decisions.

75. Ltr, Dr. A. H. Flax, ASAF/R&D, to C/S, USAF, 13 Oct 66, subj: SLV-3A Launch Vehicle Requirements, SAFSS files; msg, Dir/NRO staff to Dir SP, 24 Oct 66; msg, SP to SS, 16 Nov 66.

76. Rpt, "A Review of the Gambit-3 Program," Feb 68, inSAFSS Proj Ofc files.

77. Min NRO ExCom Mtg, 23 Nov 66.

78. Min NRO ExCom Mtg, 17 Aug 66.

79. Memo, A. H. Flax, DNRO, to DepSecDef, 20 Sep 66, subj: The DNRO Recommended FY68 Budget for the National Reconnaissance Program.

80. Min, NRO ExCom Mtg, 23 Nov 66.

81. Memo, MGen J. T. Stewart, Dir/NRO staff, to DNRO, 30 Jun 67, no subj, NRO files.

82. Memo, James Reber, Sec NRP ExCom, to DNRO, 9 Dec 66, subj: Agenda; Min, NRP ExCom Mtg, 16 Dec 66.

83. Msgs, 20 Jan 67; 14 Feb 67.

84. Memo, BGen R. A. Berg, Dir/NRO staff, to MGen J. T. Stewart, Ofc, 23 Feb 67, subj: Rapid Recovery Capability from a Disaster for SLC-4W and SLC-4E.

85. QPR, 31 Mar, 30 Jun, 30 Sep 67.

86. BGen R. A. Berg, Dir/NRO Staff, to DNRO, 25 Jul 67, no subj.

87. Draft paper, Col. C. L. Battle, SAFSS, no date (1967), subj: High Resolution Photography; memo, BGen R. A. Berg, Dir/NRO staff, to Battle, 22 Aug 67, no subj.

88. Memo, BGen R. A. Berg, Dir / NRO Staff, to A. H. Flax, DNRO, 1 Sep 67, no subj.

89. QPR, 30 Sep 67.

90. Memo, A. H. Flax, DNRO, to C. Vance, DepSecDef, 6 Jul 67, no subj.

91. Mission Summary No. 7, 15 Sep 67.

92. Msg, 26 Oct 67.

93. QPR, 31 Mar 67, and 31 Dec 67.

94. Rpt, Eastman Kodak Co, 30 Dec 69.

95. Memo, BGen R. A. Berg, Dir/NRO staff, to C. A. Sorrels, BoB, 30 Dec 68, subj: Resolution Related Improvements, GAMBIT; Rpt, Program 110 Status Book, Section 6, "Hardware Description and Capabilities," in Gambit office files.

96. Ltr. Chin, USIB, to Sec Def, 4 Apr 68, with atch, New Coverage Requirements, 27 Mar 68.

97. Memo, A. H. Flax, DNRO, to Chmn, USIB, 10 Apr 68 (draft), subj: Gambit Mission Schedule.

98. Launch Summary, " in Gambit office files.

99. Mission Summary No. 7, 15 Sep 67; QPR, 30 Sep 67.

100. Msg, 16 Aug 67.

101. Mission Summary No. 7, 15 Sep 67.

102. "Summary of Series 4300 Operational Missions," (referred to hereafter as "4300 Summary"), prep by SAFSP; Gambit Office "Program 110 Status Book" (referred to hereafter as "110 Status Book"). There is an unresolved ambiguity affecting resolution data in the records of that mission. The Gambit Office number is used in the text, while Operations shows the result to be because there were so many ways of measuring resolution, and so many points at which provisional measurements were made, that confusion is not surprising.

103. Msgs, 19 Sep 67; msg, Wir 2 Oct 67; Mission Summary No. 8, 25 Oct 67; msg, , 2. Oct 67.

104. Msg, 28 Aug 67; Mission Summary No. 8, 2.5 Oct 67; Analysis of Gambit (110) Project, Atch 3.

105. Mission Summary No. 9, 27 Nov 67; the original launch date had been 10 Oct. In turn it became 17 Oct, 20 Oct, and L4 Oct; see msgs 14 Sep 67; I- 4 Oct 67.

106. Msgs, 23 Oct 67; 25 Oct 67.

107. Msg, 6 Nov 67; memo, King to McLucas, 2.8 Apr 70, Atch 3; Addendum to EKC Evaluation, Flight No. 9, 2 Jan 68.

108. Min, of NRP ExCom Mtg, 17 Nov 67.

109. (Text Reference was Redacted) Memo, MGen J. M. Reynolds, USAF (Ret), Office of the Dir, CIA, to BGen R. A. Berg, Dir, NRO staff, 15 Sep 67; subj: Utilization of the KH-7 and KH-8 Reconnaissance Systems for Mapping, Charting and Geodesy Requirements, in DNRO files.

110. Memo, BGen R. A. Berg, Dir/NRO Staff, to A. H. Flax, DNRO, 26 Sep 67, no subj. (Flax's marginal note in reply is quoted here); memo, Ch Photo Br NRO, tau to Flax, 26 Sep 67, subj: Satellite Coverage of 26 Sep 67.

111. Memo, BGen R. A. Berg, Dir/NRO staff, to CIA, 14 Dec 67, subj: GAMBIT and GAMBIT-Cubed Film Expenditure Against Mapping and Charting Requirements.

112. Mission Summary No. 10, 22 Jan 68.

113. QPR, 31 Dec 67.

114. Mission Summary No. 11, 6 Mar 68; msg, 16 Jan 68.

115. Msg, 18 Jan 68.

116. Msg, 29 Jan 68; msg, Mission Summary No. 11, 6 Mar 68.

117. Memo, King to McLucas, 28 Apr 70.

118. Msg, 13 Mar 68.

119. Mission Summary No. 12, 1 May 68.

120. Program 110 Launch Summary, 30 Jan 68;

121. Memo, BGen R. A. Berg, Dir/NRO staff, to Dir, Join Reconnaissance Center, 8 Apr 68, subj: Satellite Reconnaissance Data.

122. Msg, 9 Apr 68.

123. Msg, 17 Apr 68; Mission Summary No. 13, 11 Jun 68.

124. Msg, 6 Jun 68.

125. Msg, 5 Jun 68; Mission Summary No. 14, 17 Jul 68.

126. Msg, 12 Jul 68. This message is also congratulatory, pointing out that the mission had exceeded in intelligence take, which had been a record to that date. Of particular note, however, was the exceptional ratio of read-out to programmed targets.

127. Msg, 17 Jul 68.

128. Msg, 14 Oct 68; Mission Summary No. 15, 7 Oct 68.

129. Ltr, BGen R. A. Berg, Dir/NRO Staff, to Dr. A. H. Flax, DNRO, 14 Nov 68, no subj.

130. Ltr, Dr. A. H. Flax, DNRO, to VP and GenMgr, EKC, Apparatus Div, 31 Oct 68, no subj.

131. *(Text Reference was Redacted)* 4300 Mission Summary, memo, BGen R. A. Berg, Dir, NRO Staff, to Dir, CIA Reconnaissance Programs (draft), 12 Feb 69, subj: Special Report on Current Soviet Anti-Earth Satellite Capabilities.

132. *(Text Reference was Redacted)* Memo, BGen R. A. Berg, Dir/NRO staff, to Analysis Branch (SAFSS), 3 Feb 69, subj: Mission 4319 memo, Pa to Berg, same subject, 4 Feb 69; memo, Berg to Dr. A. H. Flax, DNRO, 7 Jan 69, subj:

133. Mission Summary Numbers 16 to 22; Program 110 Launch Summary, 4300 Summary, Analysis of Gambit (110) Project; Quarterly Program Reviews during the period Sep 68, Sep 69.

134. The discussion of the flight program which follows is taken principally from a tabular history of the program which is maintained in the Program 110 Status Book, a working document of the Gambit Office, SAFSP. Additional sources are cited individually where used.

135. SAFSP, "Analysis of Gambit Project Missions 23 through 36," p 6, Gambit office files (hereafter cited as "Analysis of Gambit Block II"); MFR, LCol R. J. Kiefer, NRO staff, subj: Vehicle Atmospheric Survivability Test (VAST-Phase II), dtd 17 Sep 71, NRO staff files, SAFSS.

136. Memo, LCol F. L. Hofmann, NRO Staff to Dr. J. L. McLucas, DNRO, 25 Oct 72, subj: Recovery of Space Objects; msg, H. Cohen, SAFSP (SP-3) to McLucas, 25 Oct 72; interview, H. Cohen by R. Perry, 21 Dec 72.

137. QPR, 30 Jun 72.

138. Analysis of Gambit, Block II, pp 3, 11.

139. Msg, BGen L. Allen, Dir, SAFSP, to Dr. J. McLucas, DNRO, 7 Oct 70.

140. Analysis of Gambit (110) Program.

141. QPR, 30 Jun 70.

142. Analysis of Gambit, Block II, p 7.

143. Program 110 Launch Summary.

144. Memo, J. L. McLucas, DNRO, to SecDef, 18 Dec 72, subj: Taking Stock of the National Reconnaissance Program; memo, McLucas to SecDef, 21 Dec 72, subj: Taking Stock, both in SAFSS files.

145. *(Text Reference was Redacted)* Memo, BGen D. D. Bradburn, Dir/SP, to Dr. J. McLucas, DNRO, 20 Mar 73, subj: Gambit Mission Summary (4337).

146. Memo, BGen D. D. Bradburn, Dir/SP, to Dr. J. McLucas, DNRO, 23 Jul 73, subj: Gambit Mission Summary (4338).

147. *(Text Reference was Redacted)* MFR, Actg Chief, Photo Sci Studies Br, NPIC, 28 Jun 73, subj: NPIC Support to Project Skylab, in SAFSS files.

148. *(Text Reference was Redacted)* MFR, LtCol F. L. Hofmann, NRO staff, 15 Jun 73, notes by M. Krueger, NASA, 17 May 73, in NRO files; msg SOC to Dir/SP (BGen D. D. Bradburn), 17 May 73; msg, Bradburn to J. L. McLucas, DNRO, 21 May 73; msg, SAFSP to L. C. Dirks, CIA, 18 May 73; MFR, 28 Jun 73; interviews, Hofmann by R. Perry, 20 Jun 73, H. C. Cohen, SAFSP, by Perry, 6 Jun 73.

149. Program 110 Status Book,

150. *(Text Reference was Redacted)* Rpt, Analysis of G(110), 28 Apr 70, prep by Gambit Program Office; memo, Col. E. Sweeney, Dir/NRO staff, to Dr. J. L. McLucas, DNRO, 15 May 70, no subj, with unsigned memo by 14 May 70; memo, McLucas to DepSecDef, 1 Jul 70, subj: NRP Funds Available for Withdrawal; rpt, DNRO Report to NRP ExCom on Fiscal Year 1970 Status and Fiscal Year 1971 Program, 15 Jul 70.

Center for the Study of National Reconnaissance Classics

A History of Satellite Reconnaissance
The Perry Hexagon History

Robert L. Perry
November 1973

Hexagon Preface

This portion of *A History of Satellite Reconnaissance* covers the period of Hexagon gestation before April 1966 as well as the development and early operational missions of that system. At the time this preface was written, in November 1973, the agreed terminal point was July 1973. Therefore nothing that relates to Hexagon mission 1206 (the sixth flight) or subsequent operations is detailed here, and plans for improvements are discussed only as they existed in July 1973. It seems reasonable to assume that at some later time the subsequent flight and developmental history of the system will be completed, but that must for the moment be treated as conjecture rather than promise.

The author's research for this volume was supported by Robert A. Butler, at the time of writing a consultant with Technology Service Corporation, of Santa Monica, California. The history was prepared under terms of a contract between the Directorate of Special Projects (Program A) of the National Reconnaissance Office and Technology Service Corporation.

As detailed in the following pages, Hexagon was the outgrowth of effort undertaken in two earlier pseudo-program enterprises known as Fulcrum and S-2. Both have been treated here in somewhat greater detail than might ordinarily be warranted, given that Hexagon, as eventually operated, was strikingly different from its apparent predecessors. But the problems that beset Hexagon development from 1966 to 1971 were unmistakably derived, in considerable part, from the assumptions, premises, plans, schedules, and concepts that characterized those predecessor activities. As several principal officials of the sponsoring development agencies later conceded, Hexagon was prematurely advanced from engineering development to system development. Unwittingly, it became at once the most costly and the most lengthy of the several ambitious developments undertaken in the first 10 years of the National Reconnaissance Program. In the end it also became one of the most successful, and that happy outcome largely offset whatever criticisms might have been leveled at its pre-operational phases.

Because Fulcrum, as a program concept, and the Hexagon camera system as a whole were entirely CIA-managed efforts, a full history of the program should not be prepared without first reviewing CIA records. As written, this account is academically defective in that the author had no access to CIA sources. Nevertheless, the principal aspects of the total program appear to have been thoroughly documented in "Program A" records (kept in the El Segundo, California, offices of the NRO's Directorate of Special Projects) and in policy documents filed in the offices of the NRO staff (in suite 4C-1000 of the Pentagon). To the author, therefore, it seems unlikely that any subsequent expansion or enlargement of the manuscript will cause significant alteration of either the recorded sequence of events or the interpretations attached to them.

As with earlier program history contained in this set of volumes, there is no reasonable prospect of understanding the course of events in one system program without taking account of developments elsewhere in the National Reconnaissance Program. Thus from time to time it is essential to discuss events in such programs as Corona, Gambit, and Samos—and to consider in the broad the plans and policies adopted by the Director of the National Reconnaissance Program, the Director of the Central Intelligence Agency, the United States Intelligence Board, the Executive Committee for the National Reconnaissance Program, and the several other officials, boards, panels, and agencies, which influenced the establishment, growth, and conduct of Hexagon. Many of the events so mentioned have been described in greater detail in other volumes of this history: Corona, Samos, and Gambit, for instance, are the subjects of Volumes I, IIA, IIB, and IIIA of this set of reconnaissance program histories. Readers concerned about background and detail that involved those programs with Hexagon should consult those other volumes.

In the interests of avoiding repetition, most interactions between Hexagon and other programs have only been summarized here. Such summaries have been included, even if occasionally repetitious of earlier volumes, in the expectation that some readers will want to have within one set of covers reasonably complete information on Hexagon alone. This volume has therefore been constructed so that it will stand alone, without recourse to other sources, although in some instances it will be necessary to consult those other sources in order to acquire a full understanding of incidents and events mentioned casually here.

The close interaction of Hexagon and Gambit is the principal justification for making histories of those programs Volume IIIA and IIIB of the complete set. Keeping them physically separate from one another has an additional advantage: should it later prove feasible and appropriate to do so, each volume can be extended to include the later histories of those programs without forcing revision of these chapters and pages.

Finally, it is essential to acknowledge the very considerable assistance of Colonel Frank S. Buzard in providing detail and background information and in clarifying both technical and management matters that for one reason or another were either casually explained or ignored in the voluminous documentation of the Hexagon program. The source notes that follow the text do not adequately credit the comments, additions of detail, and explanations of confusing events that he provided throughout the period of background research for this volume and—most particularly—upon reviewing the initial draft. This acknowledgement must serve as the author's apology for that shortcoming of the manuscript.

Introduction and Background

Hexagon stemmed immediately from a program known as Fulcrum, which began as an Itek Corporation study initially funded by the Central Intelligence Agency in January 1964. But Fulcrum was preceded by an extended period of technological rummaging about in the requirements for a new search system—a replacement for Corona and for the failed Samos E-6. The conduct of Fulcrum and the subsequent emergence of a Hexagon program were marked by two years of variously intense controversies about requirements, schedules, technology, and organizational prerogatives.

Corona, it will be recalled, had never been intended to serve as more than an interim search system, a temporary and presumably inferior predecessor to other and more capable systems to be developed during the late 1950's and early 1960's. But by 1961 several of the planned successor reconnaissance satellite programs were in technical and financial difficulties while Corona was becoming an operationally effective and generally reliable search system with considerable potential for growth. How that potential should be exploited, and to what extent Corona might be utilized in the place of other and less attractive reconnaissance satellite systems, had become of considerable interest to the intelligence community by 1961; the composite issue of what system, if any, should eventually replace Corona, involved questions of institutional prerogatives, camera and space-vehicle technology, and national requirements for overflight photography that were not acted upon until 1966 and were not fully resolved until 1970.

Once the dual-camera, stereo-capable Corona-Mural system had been proved technically feasible, it was inevitable that a still better system based on Corona concepts and hardware would be proposed. In March 1962, the CIA endorsed an Itek proposal to develop what came to be called the M-2 search system (for Corona-Mural-2). It involved the substitution of a single 40-inch f3.5 lens and a dual-platen film system for the dual-camera Corona-Mural then in use. The estimated cost of design and manufacture seemed acceptable in that the system promised to return broad-area photography with resolution of about four or five feet for considerably less than would be expended in obtaining such performance from alternative systems then proposed or in development.

The M-2 proposal was formally presented for NRO review on 24 July 1962. Six months earlier, in December 1961, the E-5 surveillance system being developed under the aegis of the original Samos program had been severely cut back, and in July 1962 a programming error had caused the last of the E-5 recovery capsules to stabilize in a high orbit where it would remain until decay and reentry "somewhere east of Africa" more than a year later. Lanyard, a relatively inexpensive composite of E-5 camera technology and Corona vehicles, was making reasonable progress toward a scheduled first launch in December 1962, but like E-5 and Gambit, Lanyard was predominantly a surveillance system.* If Gambit were successful, there would be no need for Lanyard.

Corona, E-5, and Lanyard were Itek camera developments. The need and real potential for Corona improvement was still uncertain. E-5 had been cancelled, and Lanyard was a dubious prospect. Corona, and to some extent Lanyard, represented the only satellite reconnaissance programs under CIA control. The various Samos efforts (by 1963 reduced solely to an E-6 effort with a record of five successive mission failures and a most unpromising future), Gambit, and several radiation-sensing satellites, were under the cognizance of the NRO's Directorate of Special Projects, on the West Coast. If E-6 could be made to work, and Gambit performed as its developers anticipated, neither Itek nor the CIA could be sure of a continuing direct role in the development and operation of reconnaissance satellites.

That circumstance was well appreciated by the Department of Defense, the CIA, and all of the participating contractors. Although interagency working level relationships had been outstandingly effective during the earlier days of Corona operations, they were less so by 1963; the CIA and DOD participants in Corona were by then engaged in organizational skirmishing that was within two years to become a source of major concern to cabinet-level DOD and CIA officials.

Operating-level difficulties were paralleled by institutional conflicts at the NRO level, where they would contribute to the 1963 resignation of the CIA's designate as deputy director of the NRO (Herbert Scoville) and the later departures of an NRO director (Dr. Brockway McMillan), his CIA opposite (Dr. A. D. Wheelon), and several lesser officials. Although a variety of questions involving funding responsibilities, program management authority, and organizational prerogatives (as well as some personal differences) influenced events; a central theme in the whole period between 1962 and 1966 was the selection of a new search-mode reconnaissance satellite.

---

* E-5 and Lanyard were intended to be surveillance systems, and Gambit to be a technical intelligence system. But because only the latter became operationally available, it served as and often was characterized as a surveillance system, none other existing.

When the M-2 proposal first was formally presented to NRO program reviewers in 1962, the E-6 "successor system" originally intended to provide better search coverage capability than Corona was entering its yet-to-be-acknowledged final decline. E-6, carrying two 36-inch focal length cameras, could in several respects provide nominally better coverage than Corona, but by late 1962 a series of sequentially introduced Corona improvements had made the E-6 relatively less attractive. Then the first two attempts to operate E-6 on orbit ended in recovery failure; perhaps as important, they had been accompanied by serious camera system malfunctions. In July and August 1962, the third and fourth E-6 missions also ended in failure. In October, E-6 seemed so little promising that Major General R. E. Greer (NRO Director of Special Projects) and Dr. J. V. Charyk (then NRO director) decided to suspend plans for the purchase of operationally configured systems. The fifth E-6 sank in the Pacific in November 1962, damaged by reentry heating. Although there were indications of acceptable on-orbit camera operation before the reentry sequence began, by that time the potential advantages of E-6 over Corona-Mural had all but disappeared. The older system was returning film images with resolutions on the order of 13 feet. Even if E-6 could do better—still not at all certain—and could provide broader coverage because of greater film capacity, the Corona system had reliability attractions that E-6 seemed to lack. Notwithstanding determined efforts to diagnose and correct the defects E-6 had displayed in five successive mission failures, there was no real assurance that the system could be made to work. In January 1963, therefore, Charyk cancelled the E-6 program.[1]

The still undetermined future of Corona M-2 was clouded, during the late months of 1962, by the emergence of another Corona variant, the dual-capsule Corona-J system. Although not formally approved for development until October of that year, Corona-J had actually entered a phase of engineering design in July, with a first launch scheduled for May of 1963. (Because of problems mostly external to Corona-J, actual first launch did not occur until August 1963.) Another objection to proceeding with M-2 was the proposed development of an "improved" and re-engineered E-6 utilizing proven components in place of many troublesome elements of the original. Yet another was the lack of a stated requirement for a relatively high resolution search system, although the requirements that had warranted a 1961 start on E-6 development still remained to be satisfied.

Notwithstanding such uncertainties, the Directorate of Special Projects awarded a study contract to Eastman Kodak in January 1963 that called for examination of the high-resolution, broad-coverage mission and means of performing it. Called Valley, the project quickly focused on a large-optics system to be placed in orbit by a Titan IIIC booster. The difficulties of providing wide area coverage at such resolutions finally caused termination of that part of the study effort. The promising consequences of flying very large optics led, however, to the development of Gambit-3. Moreover, research undertaken after cancellation of the original E-6 Samos program together with the search phase studies led toward Eastman's S-2 designs of 1964.

In the Spring and early Summer of 1963, CIA reconnaissance specialists had proposed two alternatives to M-2 as candidates for the "next generation" reconnaissance satellite. One was a vehicle that could be flown covertly, that could be represented to be something other than a reconnaissance vehicle. Disagreements about the validity of and need for such a concept had been involved in Scoville's resignation in June 1963.

The second concept was even more controversial: the agency suggested the need for a system that could perform wide-area coverage at very high resolutions, the proposed requirement emerging from a series of studies conducted by CIA system analysts in early 1963. Such requirements uncertainties were passed on to the Purcell Panel, a special reconnaissance study group established by John A. McCone, Director of Central Intelligence, in the Spring of 1963.*

Perhaps surprisingly, the Purcell Panel concluded that "the natural incompatibility of wide coverage and high resolution within a given payload is becoming more acute ... as the art advances." An effort to combine the two functions in a single system "with only a modest improvement in resolution... would not be a wise investment of resources," the committee decided. Rather than to focus immediately on development of a new system, the NRO was urged to concentrate on improving the average quality of returns from Corona. The Purcell Panel made a number of specific suggestions for lines of research that promised to lead in that direction. But the panel suggested that a new system, though ultimately needed, was for the moment a lesser requirement.

---

* The "Purcell Panel," headed by E. M. Purcell, included A. F. Donovan, E. G. Fubini, R. L. Garwin, E. H. Land, D. P. Ling, A. C. Lundahl, J. G. Baker, and H. C. Yutzy—perhaps the most distinguished group of authorities on reconnaissance, space, and photography ever to be collected in one study group. Many of the "Purcell Panel" members subsequently became members of the "Land Panel," which between 1965 and 1972 operated as the principal advisor for reconnaissance matters to the President's Scientific Advisory Group and the President's Science Advisor.

The Purcell Panel report had several interesting repercussions, some of them delayed rather than immediate. One that was to become important somewhat later involved interpretation of the qualifications in the "not a wise investment" judgment. The CIA ultimately argued that the panel had endorsed development of a combined search-surveillance system with more than a modest improvement in resolution. The NRO's special projects directorate tended to emphasize the panel's view that combining high resolution with wide coverage was an exercise in natural incompatibility. But in any event, the panel plainly had refused to accept the findings of an earlier study group organized by Greer, at Charyk's direction, in April 1963. Concerned with the broad issue of what should be developed in the way of a new search system, the West Coast group (headed by Colonel Paul Heran) decided that an "improved" E-6 camera system coupled to an enlarged Corona-style recovery capsule should be developed in parallel with the proposed Itek M-2 system, the more promising of the two being produced once its superiority had been verified.

(It is worth recalling that by early 1963 the E-I, E-2, E-5 and E-6 had all been cancelled, Lanyard was in some early difficulty, while Gambit, still untested, was recovering from technical and financial troubles that in October 1962 had led to major program restructuring and the assignment of a new project head. The interest of the "Ad Hoc Group" in sponsoring parallel programs and in delaying a system choice until one or the other had demonstrated its capability for effective orbital operations becomes readily understandable in that light. So does the Purcell Panel conclusion: invest first in improved Corona quality; Corona works now. High-risk technology was in disfavor in the summer of 1963.)

The new NRO director, Dr. Brockway McMillan, ordered cancellation of M-2 work at Itek in July 1963.* Itek's efforts were to be principally focused on improving Corona product quality. To that end, General Greer's directorate made a number of specific suggestions for detail changes. CIA technical specialists in reconnaissance, now concentrated under Dr. Wheelon, concluded that the proposals were inadequate, so in October 1963 Wheelon called into being a new special study group—the Drell-Chapman Committee—"to explore the whole range of engineering and physical limitations on satellite photography…" The group, acting under a loose charter proposed by John McCone in conversation with Roswell Gilpatric (Deputy Secretary of Defense), was to be concerned not merely with Corona improvements, but also with standards and needs for new systems.

Predictably, McMillan had pronounced objections to such proceedings. He did not learn of the committee until after it had been established, he felt that its "charter" was far too broad (USIB and the NRO were nominally responsible for generating and validating requirements), and he preferred to spend NRO study funds elsewhere. McMillan also protested that Wheelon had no official role in the satellite reconnaissance program.

McCone named Wheelon his "monitor for NRO matters" three days later, and Wheelon promptly declared his intention of "…getting the CIA into the satellite business in a contributing, not just a bureaucratic way."

The most attractive prospect for new program creation still was in the search area. True, an ultra-high-resolution camera was also on the general requirements list, but it seemed several years in the future and, in any case, in 1963-1964 the surveillance concept that seemed most promising was still embryonic but certain to be an Air Force undertaking. The Drell-Chapman Committee had been critical of progress in Corona improvement; in time, that criticism was to lead to the modifications incorporated in the Corona J-3 configuration, a remarkable improvement over the original Corona-Mural. But Corona J-3 still was only a proposal, and in any case there was agreement that no Corona redesign with less scope than the M-2 undertaking could substantially improve Coronals resolution capability. Camera specialists then believed that if resolution much better than 7 or 8 feet for about half of the returned film were wanted, refinement of the original Corona would not be sufficient.**

Two events followed in close order. On 18 November 1963, the NRO's West Coast directorate contracted with Itek for general feasibility studies of a new broad-area search system and for the preliminary parametric design of such a system. That action was the somewhat delayed response to the Purcell Panel findings of June 1963. It also represented, indirectly, a continuation of search system studies undertaken on the West Coast following the cancellation of Samos E-6, earlier that year. Not quite two months later the

---

* Nonetheless, the elements of M-2 reappeared, in proposal form, at frequent intervals in later years, not finally disappearing until the availability of an operational Hexagon became reasonably certain in 1971. In subsequent incarnations the basic M-2 was given several transitory names, Corona J-4 being the best known.

** Consistent, rather than occasional, resolution of 7 to 10 feet was the Corona goal defined by the Purcell and Drell-Chapman recommendations and ultimately incorporated in the Corona J-3 program. The assumption that Corona could not generate photography with 4- to 5-foot resolutions, however much the system was modified, later proved to be incorrect. Corona J-3 ultimately provided "best resolution" of 4.5 feet.

CIA separately authorized Itek to study a remarkably similar set of problems, but specified a somewhat more ambitious design goal based on the findings of in-house CIA analyses. The CIA action was a delayed response to the Drell-Chapman Committee findings of late 1963, but it indirectly represented a continuation of the search system research approach embodied in the M-2 studies undertaken by the CIA in an effort to find a feasible improvement mode for Corona-Mural. The "West Coast Itek Study" led to S-2; the CIA-funded Itek study was the genesis of Fulcrum.[2]

The CIA's intentions were generally known to the NRO staff in December 1963, somewhat before Itek formally began work. The probability that Greer's NRO group and Wheelon's CIA group would emerge from their respective study programs with competing proposals for a new search system caused some concern among program monitors high in Department of Defense ranks. (The NRO charter then in effect included no provision for anything resembling the NRP Executive Committee of later years; the Director of the NRO was responsible directly to the Secretary of Defense, CIA participation being assured by the assignment of individuals to various NRO posts—including that of deputy director.) Earlier in 1963, Dr. Eugene G. Fubini, then serving as a senior technical advisor to the Deputy Secretary of Defense, had begun acting as a defense department spokesman in NRO matters. (In the Charyk era no such intermediary function had existed, Charyk having such an effective relationship with Secretary Robert S. McNamara that it was not needed.) Fubini had by late 1963 assumed the role of a mediator in the increasingly acrimonious contacts between McMillan and Wheelon.* In December, speaking with the implied authority of Cyrus Vance, newly appointed Deputy Secretary of Defense, Fubini proposed to McCone that the CIA assign total Corona responsibility to the NRO in return for a free hand in the development of a new search system. McMillan apparently was unaware of the offer until McCone indirectly passed it along. He rejected the compromise out of hand, insisting that the NRO had to have full authority to control Corona and that a new search system could not be arbitrarily assigned to any organization. The disagreement thus expressed persisted into 1965. McMillan's efforts to resolve the issue by obtaining directive support either from McNamara or from the White House were unavailing. The President's Foreign Intelligence Advisory Board recommended strengthening McMillan's hand during a May 1964 meeting, but the draft Presidential directive sent forward in consequence of that meeting was never signed. (The 1964 election played some part in delaying a resolution of the several controversies that afflicted the NRO, the search system requirement, and the Corona program from May through November.)

The net effect was that by January 1964 the CIA had undertaken to sponsor studies with Itek, and subsequently with Philco Corporation and other subsystem specialists, leading toward a broad-area search system called Fulcrum and the NRO's Special Projects Directorate (Program A) had begun to support a different set of studies oriented toward a different kind of search system, later called S-2. A secondary consequence was that the authority of the Director, NRO, either to control or to monitor the program of the CIA-sponsored effort had been successfully denied. McMillan certainly knew of the CIA's internal studies and of their general import. It does not appear that he learned of the existence of the funded studies by Itek and Philco until the spring of 1964, five months after their inception.[3]

EVOLUTION OF A SYSTEM

As described by Itek in June 1964, Fulcrum was to be a Titan II boosted system built around a pair of rotating 60-inch focal length cameras and a transport system for seven-inch film, the general arrangement somewhat resembling what later became Corona J-3. The scale was very different, of course (Corona carried 36-inch focal length lenses and used 70-millimeter film), but resolution was intended to range from two feet to four feet across a ground swath 360 miles wide. Carrying about 65,000 linear feet of film, the system would nominally be able to photograph more than 10 million square miles of the Earth on each mission. Although optics, camera mechanism, film transport, boost, and recovery subsystems were all "new," the film transport and recovery systems (one extremely large capsule) appeared to be the high risk items.**

---

* The principal source of CIA-NRO contention in 1963 was Corona management responsibility and authority. McMillan wanted to concentrate all Corona authority under a jointly staffed West Coast project office reporting to the Director, Program A (then Greer, later Brigadier General John L. Martin, Jr.). Wheelon, firmly supported by CIA Director John A. McCone, argued that CIA control of Corona should be enlarged rather than curtailed. The issue is discussed in greater detail in the first volume of this history.

** To that time the only film-carrying reentry bodies to be recovered by the United States were variants on the original Corona capsule of 1958 vintage. Both E-5 and E-6 had used "large" capsules intended for recovery from the sea rather than aircatch. E-5 had faults other than in its recovery system, but that too may have been faulty—no capsules were ever recovered for examination. E-6 had been cancelled solely on the evidence of five recovery failures, and two were clearly the consequence of poor capsule design. "Mercury and Gemini, NASA's man-carrying orbital systems, provided evidence that bigness was not an impossible constraint; the Mercury capsule was not unlike that tested with the E-5, for instance. But all concerned acknowledged that single "big" recovery bodies were difficult to develop, and recovery was the crucial element in any reconnaissance system of the 1960's.

S-2, as first conceived, was in some respects a simpler system than Fulcrum. Intended to have both panoramic and pointing capability, it would have better resolution in a pointing mode (three feet to four feet) than in a search mode (five to eight feet), and would cover a swath about 150 miles wide during search operations. The "early S-2" embodied new optics and camera mechanism, but would rely on the Atlas-Agena booster combination and an enlarged Gambit-style recovery vehicle. Interestingly, the first "engineering models" undertaken in the two programs were the optics of the S-2 and the film transport of the Fulcrum. Itek remained the principal Fulcrum system contractor; Greer's organization brought Kodak and Fairchild into the camera study program in September 1964 and subsequently funded space vehicle studies by both Lockheed and General Electric. Perkin-Elmer declined an invitation to bid for participation in the embryonic S-2 camera studies, but undertook some work in support of Fulcrum. While such arrangements were being made, other events occurred that were to have a considerable influence on later developments. For one, Wheelon and McCone separately proposed to McMillan and Vance respectively that CIA responsibility for both development and operation of the new search system—Fulcrum—be formally confirmed. In the meantime, the CIA provided scant data on the status of or plans for Fulcrum and forbade Fulcrum contractors to release information about their progress to any agency other than the CIA. CIA proposed to establish an internal project office initially composed of five people with Space Technology Laboratories providing technical support and serving as system integrating contractor; the principal companies concerned with Fulcrum in July 1964 were Itek, General Electric and AVCO (reentry vehicle), Lockheed (space vehicle), and STL.

That procedure and particularly the withholding of Fulcrum information from McMillan's staff, was a particular irritant to the NRO. It was not, however, unprecedented. In 1963, while questions about the desirability of starting Corona M-2 development were being considered. Greer and Charyk had attempted and very nearly carried off a similar coup. It, too, involved a search system intended to succeed or supplant Corona. When E-6 was cancelled on 31 January 1963, they very circumspectly let contracts covering the study and initial development phases of Spartan, a repackaged, largely re-engineered E-6 camera in combination with a Corona reentry capsule and Thor-Agena launch-orbit vehicles. Scoville, directing CIA reconnaissance activities at that time, had harshly questioned both the technical feasibility of a "re-engineered E-6" and the motives that underlay its proposal. Had Spartan proceeded to successful operation, it would have provided better capability than Corona. Eastman Kodak was convinced that Spartan had great growth potential—which, if true, would have negated any need for CIA development of a new search system. In the face of Scoville's opposition, Charyk in mid-February 1963 formally disapproved Spartan—but in fact both the study and the procurement of long-lead-time items needed for on-orbit tests of the proposed system continued under the cover of Program A study contracts with Eastman Kodak and General Electric. The name changed. It was listed as SP-AS-63 (Special Projects Advanced Study—1963), but in all other important respects it was Spartan.

Whether Scoville and the CIA ever learned the details of the effort remains uncertain. Special precautions were taken to prevent the untimely disclosure of "SP-AS-63" activity. All project work on the West Coast was conducted in a suite of offices provided by Eastman Kodak, located about a mile from the Program A complex. Probably no more than a dozen people of the 150 or so assigned to the West Coast establishment were aware of the activity. Even fewer were briefed in the Pentagon. No CIA people visited Eastman Kodak or that part of General Electric concerned with the "study."

The work continued until July 1963. By that time the contractors had completed the preliminary design of a system that had many of the attributes of the later S-2: wide area coverage at about five-foot resolution, dual recovery capsules, relatively simple film transport mechanism, and a variety of innovations in optics that promised consistently good returns. The replacement of Charyk by McMillan in the Spring of 1963 and the difficulty of obtaining funds to proceed from advanced study to system fabrication were, in combination, sufficient to cause abandonment of the main program in July. Eastman's private studies of improved search systems continued and certainly influenced later Eastman proposals for S-2.[4] In the event, little of the "SP-AS-63" effort was communicated to the CIA. The Agency's subsequent denial of Fulcrum information to McMillan and the NRO staff may not have been entirely motivated by the Charyk-Greer ploy of 1963, but there was implied justification for Wheelon's actions in the earlier Charyk-Greer maneuver. By the end of June 1964, when McMillan first was exposed to a full briefing on Fulcrum, the CIA concluded that preliminary studies had been sufficiently exhaustive to confirm the feasibility of the system. The request that Vance confirm the CIA's responsibility for full development of Fulcrum had been submitted. There were strong indications that the United States Intelligence Board (USIB) would shortly issue an updated search system requirement to replace those dating from 1960. On 9 July, therefore, Dr. Wheelon proposed that the NRO provide the bulk of the funds needed to support a Fulcrum development effort during fiscal year 1965. Of that total, only a portion was to be devoted to the camera system; the remainder was to go to spacecraft, booster, and system support work (including preliminary investments in the construction of a launch facility for Titan III-boosted satellites).

The timing was bad. Late in June, Dr. Fubini had been exposed to details of the Fulcrum proposal and had concluded that although it had promise it also had problems, particularly in the highly complex transport system required to deliver large quantities of film to the platens at exceedingly rapid rates. At Fubini's urging, Vance on 8 July had ruled that although the CIA could perform whatever tests were needed to determine Fulcrum feasibility, the NRO's Directorate of Special Projects should conduct comparative studies of alternative search systems. (In effect, Vance was directing continuance of both Fulcrum and S-2 work at the study and feasibility determination level, but his letter did not reach the CIA until Wheelon's request for full system funding had gone to McMillan.) By January 1965, Vance suggested, enough should have been learned about the various systems to support a rational decision on the desirability of starting full system development and, if appropriate, on the choice of a system to be developed. Given that decision, Fulcrum funding was extended, roughly 20 percent of the sum Wheelon had requested.[5]

The various studies of 1963-1964 and the generous investment in pre-design research to that time encouraged the July 1964 statement of a new and formal search system requirement. Issued under the imprimatur of the United States Intelligence Board on 29 July, it called for a single-capability search-surveillance system with the area coverage equivalence of Corona at resolutions equal to those provided by Gambit. Another system was wanted that would permit interpretation of details at the resolution level with Gambit-scope swath widths.[6] Gambit-3 would satisfy the second of those requirements; Fulcrum, as then proposed, came closer to the terms of the first requirement than did the S-2 concept of mid-1964. The requirement was not obviously the product of any single faction in the intelligence community, nor was the coincident statement of a Fulcrum-oriented requirement and a Gambit-3 requirement merely an expression of an effort to provide continuing work for both the CIA and the NRO's Special Projects directorate. The USIB had taken account of such as the Purcell, Drell-Chapman, and Land Panel studies, the comparison of M-2 and "improved E-6" potential, and several lesser analyses. And even though Fulcrum seemed nearer the new requirement than S-2, neither of the proposed systems represented a fully satisfactory solution.

While the CIA-managed effort continued, chiefly under contract to Itek but also with Philco and Perkin-Elmer, the West Coast group was devoting equivalent attention to camera system studies being prepared by Itek, Eastman Kodak, and Fairchild. General Electric and Lockheed were performing space vehicle and reentry system research for both CIA and NRO sponsors. It seemed inevitable that some version of the solid-rocket augmented Titan III would serve as the boost vehicle, whatever the final system configuration. Of the several contractors involved in some aspect of camera system design, Eastman seemed to the S-2 program office to have the most promising concept. The CIA clearly favored Itek's approach (which incorporated an optical bar system sponsored by the CIA's in-house lens specialists).

The relatively even tenor of development in parallel was rudely disturbed in February 1965; Itek abruptly renounced any intention of continuing Fulcrum work, advising both the CIA and the NRO that the company would forego any further development work on observation satellites rather than pursue the Fulcrum task as then defined. The decision was motivated by Itek's continuing disagreements with the CIA's technical monitors and the Agency's insistence that Itek defer to Agency specialists in technical matters.

Wheelon concluded that Itek's action had been prompted, or at least supported, by the NRO staff and that Itek had in effect been promised the S-2 contract in return for withdrawing from CIA-supported Fulcrum development. In fact, the NRO staff and McMillan were quite as surprised by Itek's action as were CIA officials; McMillan conscientiously advised Itek that the NRO evaluations of S-2 progress to that time showed the Eastman design to be the most attractive. McMillan had scant knowledge of Fulcrum's status at the time Itek withdrew, having received no written reports on the program since August 1964 and only sketchy verbal summaries. Nevertheless, because S-2 seemed to be proceeding nicely and the withdrawal of the chief Fulcrum design contractor could not but confuse and delay Fulcrum progress, it seemed likely that in any near-term comparison of system proposals leading to a system selection, the Eastman S-2 design would win easily.[7] The Itek affair had extensive and unexpected consequences. Perhaps most important, it exacerbated the already disharmonious relationships between the CIA and the NRO and sharpened the existing antagonism between McMillan and Wheelon. Perkin-Elmer, rather than Itek, became the principal Fulcrum camera system contractor. And, as McMillan had predicted, when the S-2 project office was obliged to designate a preferred agent for S-2 development, in May 1965, Eastman got the nod. But in the end expectations that the development of a new search system would proceed from exploratory development status to system development in 1965 proved optimistic. Although McMillan approved a plan on S-2 development in fiscal year 1966, in the event expenditures were limited to a rate somewhat below a year pending a decision on the start of the system selection process. Fulcrum funding was concurrently reduced to about the same level. For practical purposes, the effect of the Itek affair had been to delay any decision on starting development of a new search and surveillance system. Approval of that start required the concerted support of the Director of Central Intelligence and the Deputy Secretary of Defense. On 24 June 1965, McMillan advised Brigadier General John L. Martin, Greer's successor as Director of Special Projects (Program A) for the NRO that no agreement on a system approval process had emerged from DOD-CIA meetings and that none could be immediately expected.[8] Vice Admiral W. F. Raborn, who had succeeded McCone as CIA director in April, proposed to Vance in June that no action be taken on the selection of a new search system until the basic issue of NRO reorganization had been resolved. The NRO charter of 1963 was by mid-1965 being honored chiefly in the breach. Extensive readjustments of responsibility and authority in program management" funding control, operation of on-orbit satellites, and the program decision process had been proposed in the interim. But however sweeping the reorganization, it was unlikely to result in a working relationship that could accommodate both Wheelon and McMillan. As early as February 1965, a week before the Itek affair, the deputy NRO director had resigned in frustration; a senior CIA employee assigned to the NRO, he found himself so thoroughly distrusted by both staffs that he was almost totally ineffective. The S-2 and Fulcrum project groups had little direct interaction, but they were bitter competitors for funds and held divergent views on how the search system requirement should be satisfied.

Raborn's intransigence on the search system issue, the definition of a new NRO charter without inputs from the NRO, and the virtual collapse of communications between McMillan and Wheelon, the principal managers of the National Reconnaissance Program, had their inevitable effect early in July. McMillan privately advised the NRO staff that he planned to resign his post and return to private industry. His decision apparently was precipitated by the failure of a final effort to force a decision to develop the Eastman S-2 system, keeping either Itek or Perkin-Elmer as a supporting contractor. Raborn balked, and was backed by the Land Panel's judgment* that as yet insufficient data were available to support the selection of a single search system for intensive development.[9]

Although McMillian did not officially depart until 30 September, his chosen successor, Dr. Alexander H. Flax, Assistant Secretary of the Air Force (Research and Development), began to act as NRO director in July, formally exercising authority in McMillan's absence and informally monitoring NRO affairs throughout the transition period.** On 11 August 1965, the NRO charter of 1963 was supplanted by a new document that significantly altered earlier arrangements. The chief innovation was the creation of a three-member Executive Committee for the National Reconnaissance Program, composed of the Deputy Secretary of Defense, the Director of Central Intelligence, and the President's Science Advisor. The NRO director was to be a non-voting member. The committee acquired much of the executive authority previously assigned to (though not always exercised by) the NRO Director, including program and budget approval. If the NRO Director had until then possessed the authority to select and fund a new search and surveillance satellite system program, that was no longer the case. The NRP Executive Committee would thereafter make such decisions; the NRO director would oversee their execution.[10]

The program proposal that went to the Land Panel late in July from McMillan was paralleled by a program summary prepared by the Fulcrum project group. After having weighed the evidence, the Land Panel advised Dr. Hornig that "there is no technical basis for selecting for development at this time one system over the other, nor does the Panel see any urgency for making a selection now rather than, say, three months from now." Hornig advised Vance, therefore, that work on all three systems (Itek and Eastman on S-2, Perkin-Elmer on Fulcrum) should be continued at about the same rate for at least three additional months "in order to better define the advantages and disadvantages of each system." Thus, Hornig hoped, it might be

---

* The Land Panel, headed by Dr. Edwin Land, was created at the direction of the Special Assistant to the President for Science and Technology, Dr. Donald F. Hornig, early in July 1965. Its charter extended to "an overview of the NRP," but initially it was concerned with the technology of, requirements for and status of search and search-surveillance systems in development or proposed for development. The group first met on 21 July 1965 and continued to meet at irregular intervals until President Nixon abolished the office of science advisor in early 1973. The panel provided specialized technical support to Hornig and his successors, operating in some respects as a counterpart (or counterweight) to the NRO and CIA technical staffs that supported the DOD and CIA members of the NRP Executive Committee. Generally, however, the Land Panel evaluated proposals, studies, and programs rather than generating them, as was the case for the CIA and NRO special staff groups.
** Among other personnel changes in the satellite reconnaissance program in the late months of 1965 were Major General Robert E. Greer's retirement, in July, and Dr. Albert D. Wheelon's resignation informally announced in October.

possible to substantiate the performance claims for the various proposals.[11] Vance subsequently ruled that in the interim all effort was to be concentrated on the camera systems, which meant cessation of work on satellite vehicles, boosters, reentry capsules, and associated subsystems. That was decidedly awkward for both Fulcrum and S-2 managers, because in the early months of 1965 quite extensive preparations for full-scale development had included letting contracts of one sort or another with Lockheed, General Electric, and Martin. For the S-2, a Lockheed-General Electric competition was pending, while for Fulcrum a GE orbital vehicle and an AVCO reentry vehicle had tentatively been selected.[12]

The NRO-preferred configuration of S-2 in early August provided for a four-bucket recovery system (with potential growth to a six-bucket design) associated with a camera capable of providing three-foot resolution (at nadir) from an altitude of 120 miles. The payload would satisfy both search and surveillance coverage requirements if launched at a rate of six to nine systems per year. Carrying 1000 pounds of primary film (and 63 pounds of film for a stellar-indexing camera), S-2 would have a length of 50 feet, a diameter of 7.5 feet, and an on-orbit weight of 12,000 pounds for a 25-day mission. The incorporation of a supplemental crisis reconnaissance capability, as suggested by the Land Panel and the United States Intelligence Board, permitted complete access to any area of the earth between 20° North and 20° South latitude every five days.[13]

Compliance with Vance's instructions meant stopping General Electric's work on satellite control and reentry vehicles and confining Eastman's level of effort to that scheduled for August, actions that were taken early in September. The difficulties thus created were compounded by a special problem involving Eastman Kodak. That concern was then producing Gambit-I payloads, developing and building initial lots of Gambit-3 payloads, building a Lunar Survey payload for NASA under NRO cognizance,* and developing the S-2 payload. Although the decision had not yet been announced in mid-August; there was no practical way for Eastman to proceed with both Gambits, the Lunar Survey payload, and the S-2. Something had to give. McMillan's solution was to propose transfer of the Eastman S-2 design to Itek, with Itek also continuing development of the second preference S-2 camera already in process. Although complex, the transfer was not unprecedented, Itek's original Fulcrum camera design having been shifted to Perkin-Elmer in the aftermath of the February 1965 dispute between Itek and the CIA.

McMillan's proposal went to Secretary of Defense Robert S. McNamara on 30 August; on 22 September McNamara authorized termination of the Eastman S-2 activity and its transfer to Itek for design completion. Finally, there emerged a clear understanding that three camera designs were to be competitively evaluated for selection as the new search surveillance system: the Perkin-Elmer (Fulcrum) proposal, the Eastman S-2 design (generally known thereafter as the Itek/EK proposal), and the backup Itek design (usually identified as the "pancake" proposal, a term generally descriptive of the optical mirror layout preferred by Itek).[14]

Between February 1965, when the Itek-CIA disagreement suddenly flared, and October of that year, when Flax officially succeeded McMillan as Director of the National Reconnaissance Office, virtually every aspect of the search-surveillance system program had radically changed. The Land Panel and the NRP Executive Committee had come into being; both were to be dominant influences in the eventual selection of a design and a system contractor. McCone, McMillan, Wheelon, Greer, and several lesser figures in the S-2 and Fulcrum programs had left government service or moved to assignments remote from satellite reconnaissance. Perkin-Elmer had become the principal Fulcrum system developer, replacing Itek (and working more intently on the inherited Itek-Fulcrum design than on the original Perkin-Elmer design for Fulcrum), while Itek had acquired custody of the NRO-preferred S-2 system originated by Eastman (and was proceeding also with the Itek-pancake design that represented a backup for the favored Eastman S-2 proposal). Work on satellite and recovery vehicles, boosters, and supporting subsystems had largely ceased in September after having earlier advanced to the preliminary selection of design and development contractors.

On 6 October 1965, the Executive Committee for the National Reconnaissance Program held its initial meeting. The first order of business was the search-surveillance system. Colonel David L. Carter, for the NRO, and L. C. Dirks, for the CIA, briefed the committee on the three design proposals then being funded. (Until September there had been four, Perkin-Elmer had been working both on the design transferred from Itek and an alternative Perkin-Elmer design dating from the time when that company was the CIA backup for the Itek-Fulcrum design.) Although

---

* The NRO was involved in the Lunar Survey program because the readout camera being carried was a modest improvement of the Samos E-1 camera of 1960. Use of the E-1 camera and readout system was an economical means of performing the survey mission, the alternative being to develop a comparable camera system using NASA funds. In order to keep the nature and capability of earlier reconnaissance camera development secret, however, it was necessary to provide the E-1 through clandestine channels—which meant NRO control of the production process.

both suggested that proposals would be ready for evaluation by December 1965, there were indications that no competition could begin until sometime early in the following year.

Dr. Flax, charged by McNamara and Vance with reconciling the differences among the principals in the search-surveillance system controversy, presented to the Committee a comprehensive plan for proceeding toward system selection in an orderly fashion, one that would overcome the earlier tendency to use Fulcrum and S-2 as devices in an institutional squabble. Flax had early concluded that the requirement approved by the USIB the preceding year was inappropriate in that it specified technical capability rather than an intelligence objective. He proposed, therefore, to create a technical task group composed of representatives from the CIA (Fulcrum) and Special Projects (S-2) elements of the NRO. The task group, he suggested, would "prepare a statement of system operational requirement... recommend the selection of a system configuration... formulate plans for contractor selection, and ... recommend a program plan including a schedule." Flax also advised the Committee that he intended to establish a separate task group to "define the project management structure"—which meant, in practical terms, to decide what roles the CIA and Special Projects groups would play in the eventual development of the chosen system.

Flax had prepared his ground carefully. None of the Committee principals was surprised by the carefully constructed proposal for proceeding. All had seen the material beforehand. Without much discussion, the Executive Committee endorsed the Flax plan and for the first time in two years the search-surveillance program had reasonable coherence.[15]

During its second meeting, in mid-November, the Executive Committee turned its chief attention to the many other problems of national reconnaissance. The search system requirement received brief but pointed attention, the NRO's Comptroller reported somewhat ominously that the Bureau of the Budget might well take "an adverse view" of the development proposal on grounds of cost. Cyrus Vance, the chairman, asked for a formal statement of the Bureau's views—particularly relevant because, owing to the various delays in the search system program, it now appeared that Corona operations would have to be extended for at least a year past the point at which the new system had been earlier scheduled to enter service. One of the interactive complications was the necessity of diverting to the procurement of additional Corona systems some of the funds earlier planned for allocation to search system development.[16]

In the meantime, Flax had issued instructions for the deliberate evaluation of search-surveillance system proposals. He named the chief of the NRO staff, Brigadier General J. T. Stewart, to chair a management evaluations committee that included John McMahon of the CIA and Colonel Paul Heran of the NRO's Directorate of Special Projects. Carter, Dirks, and Colonel W. G. King (NRO Special Projects) were appointed to a technical task definition group. With interesting promptitude, Carter issued a preliminary paper describing the search requirement and the plan for system development. (Both had long been in preparation, of course.) The concept included use of a Titan IIIC booster (including two-or three-segment strap-on solid rockets for augmentation) capable of placing 13,000 pounds of payload in orbit; a satellite vehicle consisting of an orbital control module, a sensor module, and recovery vehicles (two reentry vehicles were suggested); and first launch 28 months "after development go-ahead." A discussion of the rationale for a two-bucket system provided some insight into the problems the new system would confront on the way to design approval: in the judgment of Carter's group, a two-vehicle configuration represented the best compromise of reliability and cost, although four or more reentry vehicles would provide a crisis reconnaissance capability only marginally present in the two-vehicle configuration. In the group's opinion, development of a three- or four-vehicle configuration would prove troublesome; Corona had provided experience in dual-reentry-vehicle operations, but there was no background for the complex cut-and-splice operations that would be required if more than two buckets were used. Finally, Carter's group maintained, "the severe weight and cost penalties of three or more RV's argue against compromising the primary mission and configurations for the crisis role."[17]

There was no question in anybody's mind that the camera system would be the pacing item in the development. It was with some dismay, therefore, that Martin and Flax learned late in October that Itek did not propose to complete a variety of essential tests, calibration efforts, and technical analyses until late July 1966. Until that work was in hand, there would be no fair basis for comparing the Itek-EK and the Itek-pancake designs. The transfer to Itek of the Eastman drawings, tools, and test data appeared to be an easy task; Eastman assistance to Itek was scheduled to continue until at least February 1966, by which time (the principals fervently hoped) Itek would be capable of carrying on independently. Flax responded, somewhat acidly, that "the Itek schedule for completion of those activities is not compatible with the anticipated decision milestone for the new search/

surveillance system." Assuming that Itek would tend to favor its original design over the less familiar EK design, Flax instructed Martin that unless Itek agreed to push both designs to evaluation readiness quickly, "we must...consider another course of action in this regard."*

General Martin assigned to Colonel Heran the delicate task of inducing Itek to agree to complete the work necessary to permit evaluation of the three principal systems by 3 January 1966. After extended discussion with Itek officials, Heran obtained the necessary commitment, but he cautioned that owing to the short period left for completion of the scheduled work it was likely that evaluators would have less confidence in an Itek-EK design proposal than in the Itek-pancake design proposal. In passing Heran's findings to Flax, Martin urged that an additional period be provided for equalizing the confidence in the two designs, so that both Itek bids would be honestly competitive with the Perkin-Elmer submission.

Flax accepted the altered schedule, and Itek's assurances of conscientious effort on both the Itek-EK and Itek-pancake designs, but he was in no position to extend the period of preliminary design past that earlier specified. He insisted that by January 1966 the three designs be available for competitive evaluation, promising that evaluators would make the necessary allowances for status differences.[18] Itek reluctantly acceded to the conditions, and on 22 November the formal transfer of the EK design to Itek custody received Martin's endorsement.[19]

In the event, it was April 1966 rather than January before the several search-surveillance system proposals were eligible for the transition to a formal competition stage. The EK design—paper and hardware—was not fully in Itek's custody until mid-January; several intervening reviews of camera system design status in December 1965 and January and March 1966 indicated that Itek's ability to cope with the EK design was developing slowly.

For practical purposes, Colonel Carter's task force spent most of its time working out the details of a Request for Proposal to be issued to Itek and Perkin-Elmer when all else was ready. The earlier rivalry between Fulcrum and S-2 approaches had not vanished, even though diminished by Flax's skillful assignment of responsibility to special interagency task forces. The CIA draft version of the Request for Proposal, for instance, called for inclusion of what was, in Carter's opinion, "the most optimistic [schedule] which could be envisioned" and provided for holding the formal pre-proposal briefing some two weeks before Itek would have completed its effort to become fully conversant with the transferred EK S-2 design. But by February Itek was capably briefing such groups as the Land Panel on the status and prospects of both designs, and by late March Flax had concluded that nothing was to be gained by further delaying the start of a formal competition.[20]

The main elements of Flax's proposal were a plan for source selection and a management plan. For the first of those, little that was controversial remained for decision, and in other than a casual way the NRP Executive Committee did not look into its details. The management plan, however, specified the organizational arrangement to be honored during the development of the system and thus encompassed all of the highly controversial aspects of CIA-NRO relationships that had troubled the National Reconnaissance Program for more than three years. Even in its draft form, as circulated for comment, it had evoked strong reactions from both CIA and NRO spokesmen. The original proposal, as worked out in advance of the 15 October 1965 establishment of the task group on management had represented a skillful compromise of organizational prerogatives. There was no longer any doubt that the CIA would exercise responsibility for the development of whatever camera subsystem won the competition. That much had been implied in the compromise arrangements of August 1965. But whether the sensor project office would be located with the main program office on the West Coast, as Martin wanted, or would continue to operate from CIA headquarters in Langley, Virginia was argued at length, and the scope of sensor project office responsibility continued to be debated for months. (Would it extend to the surrounding spacecraft structure, to the whole of the payload-vehicle structure, or be confined merely to optics-plus-film-transport and supporting components?)

General Martin, who had been NRO staff director during much of the period when divided responsibilities and ill-defined command lines had made chaos of Corona management, argued that a combined program office was essential, that co-project-leader arrangements could never be made to work. Supported by most of the NRO staff and his own West Coast group, he held out for assigning system integrating responsibility to the principal program office and limiting the sensor project office to custody over the camera subsystem alone.

Flax eventually concluded that integration of the camera with the payload must be a System Program Office responsibility, the CIA retaining

---

* The Itek and EK approaches differed in concept as well as detail. In the judgment of S-2 program managers, the EK design was simpler, less risky (in a technical sense), more certain to appear on time, and potentially cheaper.

sensor subsystem design responsibility and the Program A group on the West Coast being totally responsible for the main vehicle structures. That Solomonian edict was one of the few of the Flax proposals that occasioned arguments during the Executive Committee meeting of 26 April 1966, where final decisions were confirmed. John J. Crowley, the CIA's principal agent for sensor development, urged the Committee to assign to the CIA full responsibility for the structure enclosing the sensor system as well as responsibility for the development, production, and integration of the stellar index camera. Crowley contended, with Admiral Raborn's backing, that so extending the CIA's responsibilities would reduce the amount of interagency interface required for program management "and thereby markedly improve the possibilities of satisfactory performance within the time limits of the program."

Only one other difference of viewpoint surfaced during the Executive Committee meeting. Dr. Flax had provided that both the Special Projects Directorate and the CIA project office were to be authorized to issue program access clearances, and that each would honor without question the need-to-know determinations of the other: The CIA asked for a veto; Flax responded that his object was "to eliminate the use of security as a means of frustrating ...legitimate access to information..." The three principals of the Executive Committee met privately and alone after the briefings and discussions had ended. Vance, the chairman, advised Flax as soon as the three-man group had completed its deliberations that the program proposal had been approved precisely as submitted.[21] What had been approved was a detailed plan for conducting competitions for sensor systems and other elements of the reconnaissance satellite and a specification of the relationships that were to characterize the subsequent period of development and system operation. What remained for near-term decision was the choice of a camera design and a contractor, after which questions of satellite vehicle design, subsystem design, contractor selection, and booster design and selection might be taken up in order. The plan of April 1966 envisaged completion of development and first launch by mid-1968—roughly two years from the date of program approval. The effort to do away with the institutional rivalry that had marked the preceding three years of search-surveillance system development extended, finally, to nomenclature. In his 22 April memorandum proposing a structure and schedule for the program, Dr. Flax had noted that the system to be developed would carry the designator Helix. That name lasted less than a week; it had unwittingly been assigned earlier to another activity. On 30 April, the CIA assigned a substitute nickname: Hexagon. Retroactively, it was introduced into the minutes of the Executive Committee meeting that signaled program approval. The names Fulcrum and S-2 that had epitomized the earlier stages of the Hexagon program disappeared. None of the many principals ever expressed regret.[22]

HEXAGON: PROGRAM ONSET TO FIRST FLIGHT

The situation of Hexagon and the pattern of program development as anticipated at the time of program go-ahead were fairly represented by the several papers Dr. Flax submitted on 22 April, and which the NRP Executive Committee approved for action during its 26 April meeting.

The camera system, universally acknowledged to be the pacing element in a highly interactive program, then consisted of three potential proposals from two contractors, Perkin-Elmer and Itek. The principal Perkin-Elmer design represented that firm's elaboration on and improvement of a Fulcrum-based conceptual approach and engineering construct originated by Itek between 1964 and early 1966. Perkin-Elmer's own favored design of the early Fulcrum era had always been considered less promising than the CIA-sponsored Itek-Fulcrum approach and was not really in competition. Itek had two designs in process, the earlier NRO-sponsored Eastman S-2 design, transferred to Itek and the native Itek S-2 design (Itek-pancake), which the NRO had earlier considered to be a prime backup to what was by April called the Itek-EK design. Flax characterized the Perkin-Elmer design as "considerably changed and improved from a prior Itek effort." Although Brockway McMillan, Flax's predecessor as NRO director, had endorsed and attempted to secure development approval for the Itek-EK design while it still was an Eastman program, and the Itek-EK design approach was generally favored by NRO special projects people over the Perkin-Elmer Hexagon proposal, Flax wisely ignored all such considerations in his 22 April resume. The major problem of the moment, as Flax saw it, was how to conduct an equitable competition among three camera designs at different stages of refinement, composed to satisfy somewhat different technical and operational requirements, and representing an amalgam of studies and engineering effort by seven different groups (General Electric, Lockheed, Itek, Perkin-Elmer, the NRO's Directorate of Special Projects, the NRO's staff, and the CIA's Directorate of Science and Technology). All three surviving design approaches were nominally capable of satisfying the 1964 requirement for Corona-scope coverage at Gambit-level resolutions (given that the Corona and Gambit capabilities of 1964 were treated as baselines—there being no real possibility that any of the optical systems proposed for Hexagon could perform at Gambit-3 resolutions). There was general agreement among USIB, NRO, and CIA authorities that what was wanted was 25-30 day orbital life with single-mission capability for stereo coverage of 20 million square miles, a stellar-indexing camera, and either two or four recovery vehicles. The probable launch vehicle was a Titan IIID class rocket with two 120-inch diameter strap-on three-segment solid rocket accessory boosters; although an alternative five-segment strap-on rocket had determined advocates. Orbit weight of about 12,000 pounds seemed reasonable, although a slightly greater weight was not unlikely, given the growth tendencies of all previous reconnaissance satellites. Flax had designed the management mode for Hexagon to comply with the provisions of the II August 1965 NRO charter and related agreements between the CIA and the Department of Defense. That essentially meant that the CIA would retain responsibility for sensor development and sensor-related activities, and the NRO's Special Projects directorate (in Los Angeles) for all else in the total program. The two agencies would, for each segment of their assigned responsibilities, provide system engineering, system integration, and management.

Given those fundamentals, Flax proposed to distribute a system operational requirement, an RFP (request for proposal) covering the sensor system, a management plan, and a schedule of planned NRO actions. Attached to the submission that went to NRP Executive Committee members on 22 April was a set of five papers that carefully explained the rationale underlying the operational requirement, the RFP, and the management plan.

Although both the CIA and NRO participants in the S-2 and Fulcrum aspects of the pre-Hexagon program had conducted competitions for the spacecraft element of the total system, and both had settled on General Electric designs, Flax proposed holding a new competition, contending that not all eligible contractors had been offered an opportunity to bid to the same requirements, and noting also that the requirements reflected in his draft system operational requirement differed in some important respects from those earlier specified.* The NRO's director urged that the recovery vehicle contracts should be recompeted for the same reasons. To arguments that recompetition was wasteful of time, Flax responded that even if the most optimistic schedule then suggested proved valid, recompetition would not delay the first launch for more than a few weeks. (He also proposed a competition for the Titan IIID strap-on solid rockets.)[23]

Implementing papers went to the CIA and NRO participants in the program on 28 April, two days after Flax received formal notification that his proposal had been approved as submitted. (Some minor points of disagreement on security arrangements remained for clarification, but that did not constitute a significant problem.) Apart from the set of papers submitted to the Executive Committee, the 28 April directives included directions for the assembly of a sensor source selection board, preliminary budget guidance, and a memorandum to the Air Force authorizing the start of a competition for the Titan IIID and the preparation of system package plans for both the Titan IIIC and IIID. (As with the spacecraft and recovery vehicles, a final decision on configuration and design of the launch vehicle still had not been made.)

Sensor source selection, the first order of business, was assigned to a board headed by L. C. Dirks of the CIA and composed of four additional members, two from the CIA and two from the Directorate of Special Projects. They were scheduled to receive formal inputs from Itek and Perkin-Elmer by 22 July.** Booster source selection was entrusted to a similarly constituted board chaired by the Titan III System Program Office. Booster proposals were due by 1 September; Flax expected contract negotiations to be completed by early November 1966.[24]

On 30 April 1966, both the Special Projects Directorate and the CIA officially established Hexagon project offices in their respective organizations. Flax confirmed the nomination of the CIA to direct sensor development and named Colonel F. S. Buzard to head the Hexagon System Program Office on the West Coast. In what was assumed to be a temporary measure, Buzard arranged to have the Hexagon office physically collocated with the existing Corona program office, sharing command of the composite organization with the Corona chief. The purpose of the arrangement was to permit Buzard to draw on the experienced Corona people to supplement his own relatively small staff resources. With the start of Hexagon development, there seemed little doubt that Corona would cease operations in the reasonably close future. Obvious advantages resided in an orderly transfer of search-system responsibility from the existing system to its successor. In the event, Hexagon became operational five years after program start, rather than two, as had originally been proposed, and the transition was much more gradual

---

* In the event, General Electric won neither the satellite vehicle nor the reentry body competition for Hexagon.

** The proposals had been in preparation since February and the technical aspects of the three principal submissions were well known to the evaluators.

than Buzard had anticipated. The consequence was that at the end of three years the core of the Hexagon office was composed of people who had varied earlier experience with Corona but who had also accumulated considerable Hexagon experience.

With the approval of a Hexagon program and assignment of sensor subsystem responsibility to the CIA, existing S-2 contracts with Itek had to be terminated. Colonel Buzard negotiated the essential contract agreements with Itek between 6 May and 23 May 1966, and on the latter date Itek formally began work preliminary to a proposal for Hexagon camera system development. With issuance of the request for proposals on 23 May, both Itek and Perkin-Elmer became contractors to the CIA's newly created Sensor Subsystem Project Office.[25]

The matter of how many film capsules Hexagon would carry became the concern of a special study group on 24 May. The CIA's earlier Fulcrum schematic had been organized around the premise of one very large recovery vehicle; the S-2 proposal had never envisaged use of fewer than two capsules—and as many as four had been urged by members of both Fulcrum and S-2 study groups at one time or another.

On 25 May, Flax authorized the creation of a source selection board for the Satellite Basic Assembly (SBA) under Buzard's direction. The board included four NRO and two CIA members. By 8 June the formal Request for Proposal had received Flax's endorsement and eight days later it went to Lockheed, General Electric, McDonnell-Douglas, North American, and Hughes. (Hughes decided against participating in the competition.) Proposals were due by 22 August, one month after the scheduled receipt of sensor system proposals.

As could have been predicted, a renewed space vehicle competition was not welcomed by General Electric, which had won both the Fulcrum and S-2 "competitions" of the previous year. H. W. Paige, general manager of GE's space program organization, protested to Flax that it was basically unfair to GE to be forced to compete a third time, given that GE had originated the concept then being competed, had twice won competitions, had a skilled but unemployed space vehicle team available (unemployed because with the transition from Gambit-1 to Gambit-3 the orbital control vehicle around which Gambit had first been designed was no longer being used), and represented the only experienced alternative to Lockheed. Flax, who was aware of the problems created by his decision to recompete the space vehicle part of Hexagon, could but point out that Hexagon was neither Fulcrum nor S-2, that conditions had changed, and that he would give consideration to GE's experience when selection board recommendations were submitted.[26]

Although the final report of the recovery vehicle study committee had not yet been prepared, Buzard's people began writing the proposal guidelines for the recovery vehicle in June 1966. Because the number of recovery vehicles had not yet been decided, three designs were specified, providing for loaded film weights of 250, 525, and 1050 pounds. On 21 June Buzard urged Dr. Flax to approve a four-bucket configuration, but Flax decided to postpone a final decision until booster configuration and weight budget were better defined. Nonetheless, on 6 July, Flax agreed to the commencement of reviews of recovery vehicle proposals and agreed to issuance of requests for proposals by 19 July. The issuance of a Request for Proposal for the Stellar Terrain camera in late August completed the formal actions needed to get Hexagon development underway, but hopes that the development itself could proceed as expeditiously were to prove unduly optimistic. Almost two years were to pass before the recovery vehicles were at last put on contract although initial estimates of first launch date for the new system postulated availability of all subsystems within 18 months of program start.[27]

On 30 August—precisely as scheduled—the sensor source selection board reported its findings to Flax. The evaluators unanimously concluded that Perkin-Elmer had the better proposal and recommended that sensor development be assigned to that contractor. The preferred design was an outgrowth of the much earlier Itek-Fulcrum approach; the loser was the Itek-EK design of S-2 vintage.

Proposals had been evaluated in two categories: technical and operational qualities, and management, production, and logistics. In a scoring system that permitted a maximum possible score of 100, Perkin-Elmer accumulated a total of 69.3 points and Itek 54.7. Although the differences could be accounted for by many details of quality and resources, the board was influenced by Itek's emphasis on design for maximum resolution on the film as against Perkin-Elmer's approach of minimizing optical errors. Itek had larger and more complex optics; Perkin-Elmer emphasized other than optical considerations. The Itek design was based on use of a 48-inch Schmidt lens system with a maximum aperture of f/2.0; the Perkin-Elmer system on a 60-inch focal length lens with an aperture of f/3.0. In order to provide the desired ground resolution capability of 2.7 feet, the Itek system would have to be flown at an altitude of 84 miles as against the 92.5 nautical mile altitude required of the Perkin-Elmer optics for the same resolution. Optical design was also a factor in the weight characteristics of the two designs. For a 30-day mission, the on-orbit spacecraft weight of a Hexagon carrying a Perkin-Elmer camera system promised to be about 1000 pounds less than the comparable weight of a spacecraft carrying the Itek camera. Although there was little doubt that a booster-spacecraft combination capable of putting the heavier system in orbit could be obtained, it was difficult to ignore the obvious advantages of a weight differential so greatly in favor of Perkin-Elmer. (Camera weight differences totaled about 700 pounds—which partly represented the effects of a difference in design approaches but was in some respects a reflection of earlier Itek difficulties in getting the weight of the Itek-EK design down to the level specified during the S-2 phase of pre-Hexagon work.)

Other considerations of sensor system evaluation had a lesser influence on the decision than such fundamentals, but were not ignored. In the opinion of the evaluation group, the Itek proposals had a "significantly larger" development risk and the production tolerances required to insure proper operation of the Itek system would be much more difficult to meet than those of the Perkin-Elmer system. Further, because of optical surface quality requirements, the larger optical surfaces of the Itek system would create greater schedule and production difficulties than would the Perkin-Elmer optics.

Neither design was fully satisfactory in a technical sense. "Numerous errors" in design and analysis were sufficiently serious to cause the source selection board to question the adequacy of the engineering teams that prepared the two proposals. Yet in the end there was no reason to believe that either contractor lacked adequate technical resources. Perkin-Elmer was given a better chance of meeting the development schedule than Itek, although both schedules were admittedly tight.

Itek had a very slight lead in the management, production, and logistic aspects of the two proposals, and had proposed a development costing less than the Perkin-Elmer bid, but Itek production costs promised to be about 10 percent greater, and that offset the attractions of lower development costs.

In the end, Perkin-Elmer's 10 percent edge in weight and resolution, the lesser complexity of the Perkin-Elmer proposal, and the apparently greater maturity of the Perkin-Elmer design accounted for the appreciable difference in the scores awarded the two competitors.

Flax approved the findings of the source selection board as submitted. On 10 October 1966, Perkin-Elmer signed a contract calling for development of the Hexagon camera subsystem.

Flax received notice of the findings of the source selection board for the satellite assembly on 26 September 1966 and during November reviewed the initial reports of the source selection boards for the recovery vehicles and the stellar-indexing camera. He accepted the recommendation that Lockheed develop the satellite but withheld approval of the start of satellite vehicle work until mid-July 1967. In retrospect, that appeared to be an error of judgment because sensor design proceeded throughout that period without needed inputs from satellite vehicle designers. Through much of the intervening time, Perkin-Elmer and the Sensor Subsystem Program Office apparently believed that General Electric had won the competition. CIA sensor program managers and their principal supporting contractor, Thompson-Ramo-Wooldridge, encouraged a variety of Perkin-Elmer design approaches inconsistent with the Lockheed vehicle concept, but generally compatible with the General Electric approach. When Flax formally authorized funded Lockheed work on the satellite vehicle in July 1967, virtually the first and most difficult order of business was to redesign several features of both the satellite vehicle and the camera subsystem which, by that time, had become incompatible. Much of the work Perkin-Elmer had completed between October 1966 and July 1967 had to be redone during the last six months of 1967 and the vehicle-cum-camera interface definition process eventually required ten months of effort instead of the three months the System Program Office had originally allocated to that task. Still, the NRO director's decision to postpone starting work on the satellite vehicle seemed sound at the time; in his role as Assistant Secretary of the Air Force for Research and Development, Flax had recently seen the MOL program suffer various misfortunes because major subsystems had prematurely begun final design stages. He wanted no comparable problems to afflict Hexagon.

Proposals for both recovery vehicles and stellar-indexing camera were returned for further work, being technically inadequate in several respects. McDonnell-Douglas eventually won the recovery vehicle competition because of a superior heat shield design (an ingenious honeycomb-base silicon coating), and Itek the stellar-index camera competition (Fairchild's competing proposal involved a relatively risky electronic imaging technique as opposed to Itek's clever but conventional film-plus-lens design), but almost another year passed before contracts covering those subsystems became effective.[28]

The possibility that Hexagon might be partly or wholly substituted for Gambit had been entertained at the time of Hexagon program approval in 1965 but the issue did not become pressing until late 1966, when the United States Intelligence Board decided to give the question formal consideration. The immediate problem was finances; if Gambit purchases could be reduced, more money would become available for Hexagon development. But the high assurance of Gambit-3 success and the considerable value of photography thus generated warranted continued procurement of Gambit-3, so the USIB endorsed that course in December 1966.[29]

On the assumption that satellite vehicle development would shortly be approved, and in light of delays in the start of work on other major subsystems, the System Program Office late in 1966 proposed and secured approval of a new target date for first launch of Hexagon: April 1969. Contract definition for the recovery vehicles, stellar-indexing camera, and five-segment (rather than the originally proposed three-segment) solid-fuel augmentation rockets for the Titan booster was not scheduled until May 1967, but none was a pacing item in the program. The money spent on Hexagon by the end of 1966 was committed to subsystems other than the sensor, and most of the cost increments associated with major subsystems remained to be defined. Geodesy requirements had yet to be specified.[30] Preliminary mapping, charting, and geodesy system studies were not completed until March 1967 (and remained contentious for another year); the number of recovery vehicles to be carried by Hexagon was not decided until May 1967.

Nevertheless, the camera subsystem continued to pace program schedules. On 3 April 1967, Patterson advised Colonel Buzard that sensor development schedule slippages made October 1969 the probable initial launch date (rather than April) and that if the camera were installed in the satellite vehicle by Lockheed instead of Perkin-Elmer a further delay to December 1969 was likely. (The System Program Office concluded that system test requirements were such as to make camera-vehicle mating under Perkin-Elmer auspices inevitable even if not wholly desirable on other grounds.) Colonel Buzard's organization formally recommended, on 5 May 1967, that Hexagon carry four prime-payload recovery vehicles. Whether a fifth should be provided to return stellar-index camera film remained uncertain for the moment, so contract award was again delayed.

The approved budget included provisions for a four-recovery vehicle configuration, an improved engine for the Titan, and the initial procurement of 10 Titan IIID's. It did not provide for development and procurement of a stellar-indexing camera, deferred pending further study.[31]

Continuing problems with the stellar-indexing camera specification were Linked to the camera's ability to provide useful mapping data, principally to the Army. During the Spring of 1967, Perkin-Elmer proposed a system (dubbed SIMEC) based on the concept of printing calibrated reseau lines on normal Hexagon panoramic photography for mapping reference. Doubts about the quality of SIMEC induced Dr. Flax to convene a joint technical evaluation committee to examine the Perkin-Elmer proposal. The committee members (from Program A, the CIA, and such other groups as the Army Mapping Service and the National Photographic Interpretation Center) were not impressed. They concluded that SIMEC could not meet the Army's requirement for 1:50,000 scale maps, that it promised to be excessively costly, and that the reseau pattern would obscure the underlying Hexagon imagery to an unacceptable degree. The committee's recommendation was to abandon efforts to incorporate mapping capability in the Hexagon panoramic camera.

Although the System Program Office had earlier concluded that a 12-inch stellar indexing camera was needed to satisfy Army mapping requirements, action to that end was not immediately feasible because of CIA objections. But in July Flax finally announced that Lockheed had won the satellite vehicle competition of the previous summer, and contractually covered work formally began. Final contracts were not signed until December 1967, however. In early August 1967, following the announcement that Lockheed had won the satellite vehicle contract, the program office created four interface working groups for (1) structural and mechanical issues, (2) tracking, telemetry and command/electrical issues, (3) test and assembly coordination, and (4) operations. The working groups subsequently induced major changes in the design of Lockheed's orbital vehicle, a new command system and a single rather than a dual vehicle shroud being two of the earliest.

Late in October 1967, General Electric contracted to deliver a development test unit and six flight-qualified Mark IV Command Programmer subsystems adapted for Hexagon. The Lockheed and General Electric contracts were the first to be signed in the Hexagon program—other than that covering Perkin-Elmer. The delay between program approval and contractual agreement was nearly 18 months. The basic problem was lack of agreement on detailed system specifications and production quantities. In November, for instance, the NRP Executive Committee reduced the initial Hexagon buy (deleting two planned reserve systems) and the Directorate of Defense Research and Engineering formally urged that a 12-inch (focal length) stellar-indexing camera be used in Hexagon instead of the earlier proposed 3-inch design. Dr. John Foster, director of the defense engineering agency argued that no other expedient could satisfy Army needs.* The cost implications were alarming, given that Hexagon was edging toward substantial price increases in several areas, but the additional weight of a 12-inch camera was yet a larger difficulty. The fundamental objection, nonetheless, was the CIA argument that Hexagon should not carry mapping equipment at all. Again, a final decision was put off.

Although a contract covering the initial lot of 10 Titan IIID boosters became effective in December 1967 (backdated to cover "black" work Martin had performed since July), problems created by the delay in starting work on the Lockheed satellite vehicle negated any progress thus implied. By the time Lockheed was legally entitled to start final design, much of the Perkin-Elmer camera system had been configured to conform to the losing General Electric spacecraft design. In particular, the Perkin-Elmer design had to be changed so that the film supply reels were oriented along the pitch axis of the vehicle rather than along the roll axis. Reconciling other aspects of the Perkin-Elmer system with the satellite vehicle forced redesign of both in December. However, the program office was finally able to let contracts for computer software, recovery parachute design and development, and communications equipment. The effect of all that was to drive up budget levels between September and December 1967. Most of the increase reflected booster purchase costs but program changes of various sorts were important contributors.

Late in 1967, the stellar-indexing camera issue again surfaced. Deputy Secretary of Defense Paul Nitze, Chairman of the NRP Executive Committee, had become receptive to John Foster's advocacy of a large-camera stellar-indexing and mapping subsystem for Hexagon. Cost-factor objections to the proposal had been countered by an Army offer to contribute toward development. Even though no new camera could be readied in time for the first flight of Hexagon, and only the Army mapping agency maintained that a large camera was essential to the satisfaction of national requirements for maps and charts, the Army's arguments, and their sponsors, proved, compelling. The Executive Committee had to accept Flax's assurances that no large-camera stellar-indexing and mapping system could be incorporated in early Hexagons, but development of that system continued and eventually the Committee agreed that it should be used in the seventh and all later launches. No formal contract was to be signed for another year—until November 1968—but Itek continued preliminary development activities in the interim.[32]

Eighteen months after program approval, Hexagon still was making only slow progress toward operational readiness. The difficulties of proceeding from conceptual design to engineering development had been sadly understated—as had the costs of that transition. Most delays had origins in delayed decisions and management disputes, but that did not diminish their effect. The CIA's reluctance to agree to software specifications and CIA efforts to acquire control of software programs caused delays in that area, for instance. Similar difficulties occurred elsewhere: a formal system performance requirements statement appeared in January 1968, after having survived a strenuous informal review the previous November, but immediately became a matter of contention between the CIA's sensor specialists and the main Hexagon program office. Lockheed finally signed a definitive contract for space vehicle development in January but was immediately obliged to propose a major vehicle redesign in order to accommodate camera system changes made since Lockheed's design had first been submitted, some 14 months earlier. Whether the camera subsystem would be mounted by Perkin-Elmer or shipped to Lockheed for installation had not yet been decided. For that matter, still unresolved questions of camera design included decisions on the film path, the kinds and quantities of test equipment, and the scope of camera system testing to be performed once the camera section had finally been passed on to Lockheed. That Hexagon would include four recovery capsules was certain, but whether the product of stellar-indexing camera operations should be returned with the last of the four or in a separate (and smaller) capsule still was a study question as late as April 1968. Because of design uncertainties, no recovery vehicle contractor had yet been chosen. (McDonnell-Douglas had an attractive proposal in a technical sense, but the cost was unacceptably higher than for GE's less appealing

---

* Although the Directorate of Defense Research and Engineering participated in general discussions of the National Reconnaissance Program at the Executive Committee level, Foster had no vote in program decisions and little influence on most. That constraint did not extend to geodesy and cartography, however. The tradition of tri-service participation in the reconnaissance effort generally gave the Navy a major role in passive electronic reconnaissance and assigned to the Army prime responsibility for mapping and charting. When the Argon program first was approved, in 1958, the Directorate of Defense Research and Engineering inherited from the Advanced Research Projects Agency both a sponsorship function and an active voice in mapping program decisions–reflected in the composition of the configuration control board for Argon. Argon had long since passed from the scene, but Army interests still were represented by the Directorate of Defense Research and Engineering whenever mapping programs were considered. Thus Foster was in one sense a spokesman for Army viewpoints. His access to and influence with the upper echelons of the Department of Defense made that an important consideration in decisions on new stellar-indexing and mapping systems.

shield concept.) Costs were rising, schedules were slipping, essential test articles remained undefined, and disagreements over management responsibilities repeatedly disrupted routine. Nevertheless, in April 1968 program managers agreed that 1 October 1970 was a reasonable first launch date (one both could accept) and made that, rather than mid-1968, the new program goal.

Resolution of stellar-indexing camera uncertainties* early in 1968 permitted the issuance of "go-ahead" letter contracts for recovery vehicle development (to McDonnell-Douglas) late in May and for stellar-indexing camera development a month later (to Itek). Formal contracts appeared on 30 September and 15 November 1968, respectively. In May, Dr. Flax settled the, who-does-what argument over camera-vehicle integration responsibilities by accepting the CIA's contention that Perkin-Elmer could do the job of installing the camera system in the vehicle assembly more effectively than could Lockheed, thus permitting disposition of several lesser questions still hinging on that fundamental issue.[33]

Fulcrum, the 1963 proposal that eventually led to Hexagon, had initially been conceived as a search system to replace Corona. Eventual approval of Hexagon development expanded that concept to include surveillance by incorporating the 1964 "Corona coverage at Gambit resolutions" statement. Between 1964 and 1968, considerable advances in reconnaissance technology had affected both Corona and Gambit; the former had become a highly cost-effective search system with remarkably good reliability and the latter a surveillance system with a demonstrated resolution capability and evident growth capability to "best resolution." capability. Several proposed camera systems with at least that resolution potential were beginning to demand attention by 1968. Further, some of the more optimistic participants in the satellite reconnaissance effort had by that time concluded that it was now feasible to undertake development of a high-resolution readout system with near-real-time capability. In the growing national uproar over the costly Indochina War, defense budgets were becoming tighter; one consequence was that the development of expensive new satellite reconnaissance systems was becoming increasingly dependent on finding the necessary money within ceiling-limited NRP budgets. Hexagon was the single most expensive item of the 1968-1970 National Reconnaissance Program.

Starting in mid-1968, therefore, and continuing for a full year, proposals for reorientation, cutback, or cancellation of Hexagon were frequent, serious, and loud. They began routinely enough in budget bureau suggestions that Hexagon program costs were excessive and that the mission Hexagon had been designed to perform could be as well performed by other, less costly systems. That entirely legitimate issue tended to get submerged in the subsequent advocacy of particular "other" systems, partly because the McNamara tradition of proposing "alternatives" had become a fixture of the system evaluation process, partly because various groups within the satellite reconnaissance community had taken to sponsoring one particular system, and partly because any decision to cancel or reduce expenditures on Hexagon could not but enhance the prospects of some other proposal for reconnaissance satellite development and operations.

The opening of Strategic Arms Limitations Talks (SALT) with the Soviet Union further complicated orderly consideration of the future of Hexagon. Progress in the arts of satellite reconnaissance had been so rapid in the mid-1960's that it was no longer essential to couple arms limitations to the on-site inspection of strategic weapons stockpiles and installations. The Soviet Union had consistently refused that concession; pre-1968 efforts to agree on means of verifying compliance with arms limitations agreements had grounded on the inspection issue. Although neither the Soviets nor the Americans were fully prepared to specify that all needed verification and inspection could be performed by means of cameras in orbit, de facto acceptance of that premise was evident after 1968.

Once the means had been agreed upon, however informally, the details became all important. On the American side (and conceivably on the Soviet side as well), the scope and detail of coverage required to confirm compliance with arms agreements were contentious issues. Most reconnaissance systems in service and in development by 1968 had been designed to provide specialized coverage of various kinds. Systems like Corona, Gambit, Hexagon, and the several passive Elint and Comint sensor satellites could undoubtedly serve the needs of verification, but none was optimized for such an application. Optimization—which implied acceptable costs, frequency of coverage, and detail of return—could well require the development of some new system or reliance on some combination of systems not previously contemplated. Crisis reconnaissance, a troublesome subordinate requirement for a decade, could well become a dominant requirement in an era of strategic arms detente. Very high resolution might

---

* The concept of a long-focal-length camera prevailed, but not until June was it possible to confirm the advisability of returning the film product of the stellar-indexing camera in its own separate recovery capsule. The Corona capsule eventually was adapted to that purpose.

be needed to detect subtle shifts in strategic posture. Emitting systems, capable of functioning in the presence of the heavy cloud cover and poor lighting conditions that characterized most Soviet strategic missile bases, could become vital. A capability for near instantaneous recurrent coverage of selected areas might be essential. All that seemed certain was that requirements for the 1980's were uncertain and that a satellite reconnaissance system (or systems) capable of verifying Soviet compliance with arms limitation agreements must be in the American inventory in the 1970's and after. The contribution Hexagon could make, and Hexagon costs thus became factors in deliberations on the long-term composition of the National Reconnaissance Program.

Such issues began to concern the NRP Executive Committee during the summer of 1968. Late in that summer, Deputy Secretary of Defense Paul Nitze, alert to the increasing costs of the Hexagon program, the remarkable new capabilities being demonstrated by other reconnaissance satellites, and the potential value of Hexagon in a SALT-agreement verification setting, instructed Dr. John Foster, Director of Defense Research and Engineering, to undertake a comprehensive evaluation of Hexagon. Similar studies had been completed and reported to the Executive Committee at intervals since 1964 (although only lately had SALT been of real concern), but most had been undertaken by one or another of the several participants in the satellite reconnaissance effort (the CIA, the NRO, NPIC, DIA, and the NSA had all participated or contributed at one time or another), and Nitze wanted a fresh and entirely independent viewpoint.[34]

Cost was in no way a new issue. But during the summer and fall of 1968 it became apparent that substantial reductions in prospective NRP budgets for fiscal years 1969 through 1973 were inevitable and that one way of offsetting them would be to cancel Hexagon. The objection, of course, was that Hexagon returns seemed essential to satisfaction of approved NRP objectives for the post-1972 period. At that point in the discussions, the Bureau of the Budget revived an earlier suggestion that the combination of Gambit-3 and an improved Corona (presumably some variant of what was generally known as the Corona J-4 proposal) would satisfy the requirement at a cost perhaps below that anticipated for Hexagon. The CIA, DIA, NPIC, and NRO responded in concert that without a complete redesign (with costs then estimated to be equal to those of completing Hexagon development), Corona could never provide search resolutions much better than about 4.5 feet—and all those agencies were agreed that search resolutions better than 3.0 feet were essential to verification of arms limitations agreements. The Bureau of the Budget rejoinder that a 1.5-foot difference in resolution could not possibly be worth what it would surely cost by 1973 had no evident effect.[35]

In November 1968 the American electorate chose Richard M. Nixon to succeed Lyndon B. Johnson as President. Nixon appointees took office in January 1969. Foster and Richard Helms, Director of Central Intelligence, were among the few senior officials to carry over from one administration to the other. Nitze was succeeded by David Packard as Deputy Secretary of Defense, and Clark Clifford, President Johnson's last Secretary of Defense, by Melvin Laird. Clifford had delegated responsibility for virtually all matters concerned with the National Reconnaissance Program to Nitze; Laird did the same for Packard, but kept closer tabs on NRP policy decisions than had Clifford. Laird's instructions from President Nixon were to reduce defense expenditures below the levels proposed by the Johnson Administration, and he did not propose to exempt the NRP from funding cutbacks. The new Director of the Bureau of the Budget, Robert P. Mayo, had received similar instructions: he found a ready advocacy of NRP funding cuts embedded in the permanent staff of the bureau.

Very shortly after taking over the budget bureau, Mayo proposed cancelling Hexagon and substituting a Corona-Gambit capability. Packard saw little merit in the idea (he had concluded that if any major reconnaissance program were to be cancelled it should be a measure that would have about the same financial effect as a Hexagon cancellation), and for the moment Mayo received no support from the White House.[36]

Late in March, Mayo again marshaled budget bureau arguments against Hexagon and carried them to the President. On 9 April 1969, President Nixon ordered Hexagon to be cancelled. The rationale of the decision was extremely complex, but in essentials it derived from the evident necessity of eliminating either Hexagon or a competing project if the fiscal 1970-71 budget was to remain in balance, the apparent overlap of capability between Hexagon and the competing project in SALT terms, the impossibility of cancelling Gambit until a replacement was operational, and the lack of any other obvious reconnaissance program candidates for cancellation. Corona was so inexpensive as compared to Hexagon that its continuation into an indefinite future would have no appreciable effect on NRP budget levels.

Whatever the reasoning behind the 9 April decision, reconsideration was immediate. At Helms' urging, the President delayed action on Hexagon cancellation for two weeks. In that interval Helms and Packard made their objections known to the President, and on 21 April Mayo reversed his original stand. The three brought Laird to their way of thinking by late April. The fundamental argument they settled on (eventually presented by Mayo) was that Hexagon would provide a much better capability for validating any arms limitation agreement. John Foster did not fully agree, but his reservations about Hexagon derived partly from the inconclusive study he had undertaken at Nitze's urging six months earlier. (*Endnote 37 redacted.*)

The June 1969 decision was conclusive, and before long was irreversible. To have cancelled Hexagon after the summer of 1969 would have decimated the national capability for search-satellite operations. Proposals for extending Corona production and even for stockpiling Coronas against some future need (which presumably could have included the failure of the Hexagon development program) gained an occasional hearing thereafter, but never again did they have high-level support. The National Aeronautics and Space Administration wanted Corona for possible use in Earth Resources Survey assignments and the Department of State urged retention of Corona capability against crisis reconnaissance needs, but NASA was unable to finance continued Corona production and State could not overcome arguments that Hexagon would outperform Corona in a crisis reconnaissance assignment. Enough Corona systems had been ordered to protect against a serious gap in coverage should Hexagon be delayed in development—which proved notably wise—and the development of a reasonably effective and not too costly Gambit modification (Higherboy) represented another hedge against delayed Hexagon availability. Both were stopgap measures, of course; by 1969 successful Hexagon operations in 1972 had become an integral of national reconnaissance policy.[38]

During the first two years after Hexagon program approval, incurred delays had largely arisen in uncertainties of program definition and design. Their effect had been to cause a significant slippage in program schedules. Although their advocates had represented both S-2 and Fulcrum to be fit for full system development by late 1965, not until the Spring of 1966 had a development start been approved, and not until 1968 were all of the essential elements of the Hexagon system under contract. Decisions on booster configuration, recovery vehicle configuration, the selection of a stellar indexing and mapping camera, and accommodation of the orbital vehicle to the changing design of the camera system had been delayed far longer than could reasonably have been anticipated. Long after, the chief CIA manager of reconnaissance program matters concluded that insufficient background research had been performed on Hexagon in advance of the decision to proceed with full-scale system development.*

After system definition had finally been completed, an event that was difficult to date but could most accurately be assigned to mid-1968, Hexagon began to encounter the sorts of engineering and test problems that had marked the development of all earlier photographic satellites. Gambit-1 had come closer than any other photosatellite to meeting its schedule, and even Gambit had demonstrated disturbing operational shortcomings during its first year of operation. Corona had nearly been cancelled after a first year of flight experience dominated by mission failures, and all other photo-satellites of the 1960's had eventually succumbed to one or another of several major flaws. Although the Hexagon program schedule made some allowances for slippages caused by unforeseen engineering difficulties, in the end they proved to be insufficient.

The lack of an agreed software program delayed progress in operational planning until March 1969, when Flax intervened to resolve disagreements between the sensor (CIA) and system (SAFSP) program offices. Another delaying element arose from a difference of opinion involving the System Program Office and the National Photographic Interpretation Center concerning the accuracy requirement for attitude determination devices. Although for a time it appeared that some redesign of the attitude sensors might be required, in the end the problem was reduced to one of data requirements, NPIC relaxing its original demands for extreme precision. Colonel Buzard later summed up the program office viewpoint with the phrase, "if a thing is not worth doing at all, why do it well?" Nevertheless, such problems hinted at real slippages to come.

The first unrecoverable slippage of any kind was acknowledged early in 1969 (while the scheduled first launch date still was 1 October 1970); Perkin-Elmer spent an unprogrammed two and one-half months of additional work in completing and testing the first qualification model of the camera-vehicle midsection assembly. The disclosure of that misadventure had

---

* Carl Duckett, the CIA's Deputy Director, Science and Technology, suggested in a 15 July 1971 discussion of probable cost growth in a proposed new system that "...in the case of HEXAGON... we had spent little money and knew very little what we were trying to do" at the time of program approval. Although only Dr. Flax and his immediate staff seem to have expressed such misgivings while Fulcrum and S-2 were being roundly endorsed by their respective sponsors in 1965-1966, that retrospective judgment seems sound. Only the camera subsystem design seems to have been reasonably well defined at the time of Hexagon program approval in April 1966, and once engineering development got well underway even that changed significantly.[39]

been preceded by a rather unsettling special review of Hexagon engineering work undertaken to the end of 1968; the review report highlighted 14 major and 28 lesser system faults that required prompt attention. Camera subsystem development costs increased in the first quarter of calendar 1969—a foretaste of much larger cost growth to come—and various slippages and redesign requirements forced the allocation of additional funds to Lockheed.[40]

The CIA's Sensor Subsystem Program Office initially reacted to word of potential slippages in camera development schedules by proposing to compress and abbreviate elements of the thermal testing program, but that expedient became inadvisable when the camera section proved to be more sensitive to minor temperature variations than had been assumed earlier.

Although to that time only about two months of unrecoverable slippages in the total Hexagon program had been positively identified, and schedules had been designed to accommodate at least that much slack, in June 1969 Dr. McLucas* assigned to his principal deputy, Dr. F. Robert Naka the task of determining the viability of the Hexagon launch schedule (which then called for first flight no later than December 1970). Naka's evaluation,** forwarded to members of the NRP Executive Committee on 20 June, contained carefully qualified expressions of caution.

In addition to evaluating the probability that Hexagon would be launched as scheduled, Naka estimated the degree of confidence the NRO should have that the first Hexagon mission would be successful, and looked at various ways of optimizing search mission products at least possible cost. An unavoidable parallel issue was whether Corona vehicles additional to those then on order should be purchased as a safeguard against a lapse in search coverage that might occur if Hexagon operations began appreciably later than December 1970.

Naka calculated a 95-percent probability for a first Hexagon launch no later than June 1971, rated at 75 percent the probability of a first launch by March 1971, and assigned a 50-percent probability to launch no later than January 1971. He concluded that about 75-percent confidence should be assigned to the possibility of mission success on the initial flight and foresaw a 95-percent probability that at least one of the first three missions would be successful. Given those odds, he suggested that the 12 Coronas programmed for launch at about two month intervals between June 1970 and July 1971 should be rescheduled to allow for at least two missions after July 1971, thus insuring a minimum overlap of Corona with Hexagon and providing some search coverage in the event of either a Hexagon slippage past June 1971 or mission failure. Given the existing uncertainties of Hexagon scheduling, Naka also cautioned that the need for more Coronas should be reassessed in December 1969.[41]

The Naka report, standing alone, was cause for mild uneasiness. Taken together with revised estimates of Hexagon costs in fiscal 1970, however, it prompted a serious Executive Committee discussion of the future of Hexagon as a system. Both Perkin-Elmer and Lockheed had advised program managers of potentially massive Hexagon cost growth—a particularly disheartening development at a time when other elements of the National Reconnaissance Program were also in financial distress. Part of the difficulty arose from the necessity of diverting defense dollars to the increasingly costly IndoChina War; another part derived from President Nixon's assignment of a high priority to the effort to develop a near-real-time readout system for reconnaissance satellite applications—the target date being 1976. The program had been cancelled and there still was no agreement on whether ultra-high resolution or readout should receive funding priority. David Packard, chairman of the Executive Committee, asked flatly on 8 August 1969 whether there was agreement in the Committee that Hexagon development should be continued. The vote was in favor of proceeding; there was no real alternative, although various substitute means for providing search coverage in the 1970's still were being examined. CIA's Carl Duckett assured Packard that costs had been brought under control and that Perkin-Elmer, the chief offender, had promised to be attentive to the need for careful control of costs. Although the system was somewhat behind schedule, the quality of systems then in test seemed quite good, Duckett added.

In the end, the Executive Committee approved the Hexagon budget for fiscal 1970 about as submitted, merely adding a caution that the National Reconnaissance Office must keep a sharp eye and a tight hand on costs.[42]

Costs were not unrelated to schedules, of course, and in the late months of 1969 schedules were becoming almost as worrisome as costs. To maintain the required pace of progress, several contractors had resorted to double and triple shifts and the extensive

---

* Dr. J. L. McLucas succeeded Flax as Director, National Reconnaissance Office, in April 1969.

** Dr. Naka signed and reported the findings as spokesman for a committee that included the CIA's sensor project office and Colonel L. S. Norman of the NRO's Directorate of Special Projects. Although preliminary findings were forwarded to the Executive Committee in June, formal reports seem not to have been prepared until September 1969.

use of overtime. Conduct of that sort was somewhat out of fashion by late 1969, at least for most defense procurements, but for Hexagon there seemed to be no useful alternative. In the development and production of many weapon systems, the schedule urgency attached to programs was largely artificial. Major systems had characteristically been delivered from one to three years late without significantly lessening total defense effectiveness. The customary response to development delays was to slip delivery schedules and to extend the in-service life of whatever was currently in the inventory rather than to trade money for time. Aircraft program schedules, for instance, could be restructured to offset cost increases in a given fiscal period and the worst consequence was to delay the availability of some system that probably need not meet whatever schedule had originally been established. Thus overtime generally was not encouraged for normal defense procurements, and multiple shifts usually were permitted only when some critical item like ammunition was in dangerously short supply.

But the Hexagon case was in quite another category. Satellite reconnaissance systems did not stay quietly in the weapons inventory; they were expended, regularly and inevitably. If Hexagon did not appear as scheduled, some provision would have to be made for obtaining substitute coverage of target areas—and in late 1969 the alternatives were alarmingly few. Hexagon overtime and multiple-shift work was necessary to meet schedules that were based on the planned expenditure of existing stocks of reconnaissance satellites, chiefly Corona systems. Corona J-3 could not offset Hexagon requirements, and by 1969 there was no reasonable possibility of developing an improved Corona in time to substitute it for Hexagon. Indeed, within a few months it would become impossible to order additional Corona J-3 systems in time to offset a major delay in Hexagon availability; the lead time for Corona was 18 to 24 months, which meant that systems ordered in December 1969 could not be delivered sooner than June 1971. The question of whether to spend money for Hexagon overtime and multi-shift operations or to keep Hexagon on a normal schedule and buy Corona vehicles (or the only other feasible option, Gambit systems configured for high altitude flight and artificial search capability thereby), was more academic than real. The Executive Committee had little choice.

Concern did not vanish, nor did the Committee lose sight of the problem. In October 1969, Dr. Naka again reviewed Hexagon status and although an indicated additional slippage of at least one month had appeared since August, he recommended that the decision on an additional Corona buy be postponed until January 1970. By January there had been no significant change, so the Committee somewhat reluctantly decided to forego the option of ordering more Corona vehicles.[43]

Dr. Naka's report did not stand alone and unsupported—or supported only by classical contractor and program office optimism. In December 1969, Brigadier General W. G. King (who in August 1969 had succeeded Major General John L. Martin, Jr., as NRO head of Program At the Directorate of Special Projects) convened a special meeting of Hexagon principals from the program office, the sensor project office, and the major contractors to reevaluate the prospect of meeting the scheduled December 1970 launch date.*

All agreed that although the schedule was getting tighter with the gradual disappearance of slack time that had earlier been provided to accommodate inevitable engineering and test difficulties, the December 1970 deadline was reasonable—but staying on schedule would require "vigorous action" by all concerned.[44]

Thermal control testing had, by late 1969, disclosed that the camera section was not immune to internal temperature variations of as much as 20 degrees (Centigrade), as had been intended. In practice, a variation of from three to seven degrees was as much as the camera could tolerate, and in a 60-foot vehicle exposed to variant sun angles the ambient internal temperature range was much larger. Heat had to be provided for part of the system, a modification that required adding both solar panel area and more electrical power. Recalibration and rework problems disarranged the combination camera-midbody tests at Perkin-Elmer in February and March 1970, causing Lockheed to substitute available satellite vehicle test sections for those originally scheduled to be so tested, but by adroit shifting about of test sections both contractors managed to stay reasonably close to the milestone schedule imposed by the December 1970 launch date. But that sort of test rescheduling caused expenditure of very nearly the last remaining reserves of slack time in the pre-launch test program. In early July 1970, Dr. McLucas was able to report to the NRP Executive Committee that notwithstanding "...the normal difficulties one can expect with major development programs," the December launch date for Hexagon still seemed achievable.[45]

---

* Nearly seven years earlier, then-Colonel King had somewhat abruptly been named to head the Gambit program office at a time when that system was in a situation of technical, financial, and schedule crisis. A decade still earlier, he had been called on to rescue the Snark missile system after it had experienced a 300 percent cost overrun, a five-year availability slippage, and a succession of incredible technical shortfalls. He had performed admirably in both assignments. (NB: General Martin's departure was a routine reassignment after seven years with the NRO.)

Unhappily, even while Dr. McLucas was assembling his report to the Executive Committee the validity of his cautious optimism was eroded by events at the Perkin-Elmer plant. On 7 July, the first flight-article twin camera assembly (P-1) suffered a catastrophic failure during low temperature chamber testing. It had been scheduled for 31 July shipment to Lockheed. The extent of damage was so great that no possibility of timely repair and recalibration could realistically be entertained. On 10 July, therefore, the sensor program office confirmed the contractor's judgment that the second sensor system (P-2), originally scheduled for 5 September shipment, had to be substituted in first-flight schedules. It was conceivable that P-2 could be qualified and shipped by 26 August, but given the earlier disappearance of virtually all remaining slack time in the flight readiness schedule; there was slight prospect of meeting the 17 December 1970 first flight target date. Indeed, Dr. Naka reported to McLucas that even if the schedule were allowed to slip by three months (into March 1971), confidence in meeting the new flight schedule would remain low. By adopting a seven-day, three-shift operation, Lockheed conceivably could complete qualification and calibration of the combined camera-vehicle midsection assembly late in September, after which arrangements for a December launch still might possibly be made, but the effort would cost more in additional funds for Lockheed and Perkin-Elmer efforts, and still the assurance of launch would be tiny.[46]

Although the full extent of the problem was not known at the time the NRP Executive Committee met on 17 July, the implications were plain. J. R. Schlesinger, then acting Deputy Director of the Office of Manpower and Budget, promptly resurrected the proposal to buy Corona systems to fill the search-system gap that seemed certain to develop if the Hexagon camera failure was symptomatic of a major defect. Dr. Naka, whose committee had recommended bypassing that option six months earlier, explained that the last chance to order Corona systems had lapsed the previous February. If Hexagon failed, and Corona launches continued at their planned rate, there would occur a lapse of about six months before new Corona systems could be delivered. At that time (July 1970), an 11-month overlap still existed—assuming that Hexagon could meet a June 1971 launch date, the worst possible contingency previously examined in detail and that at least one of the first three launches was successful in returning search photography. The decision that had to be made, Naka explained, was whether to push for an early launch so as to learn promptly what on-orbit problems Hexagon faced, or to complete a thorough sequence of tests in order to generate high confidence in flight success and accept the resulting schedule slippage.[47]

Although on the surface the potential for launching in December 1970 looked hopeless, the sensor project office held stubbornly to that goal for nearly a month after the failure of P-1. The chosen course had been to opt for an early launch rather than extended testing. Perkin-Elmer and Lockheed overtime costs were accepted as the price of the effort. But following arrival of the second camera payload (P-2) at Lockheed's Sunnyvale facility, major problems with the film transport mechanism again stalled the test program. Faulty platen functioning and film supply operations were simultaneously delaying qualification tests at Perkin-Elmer's Danbury (Connecticut) plant. The situation having degenerated so completely, the sensor project office conceded "…that they don't have a prayer of meeting the 17 December launch date."[48]

Formal acknowledgement of the inevitable launch date slippage came from General King on 15 September: "Problems principally associated with acceptance testing of the sensor subsystem…" had invalidated the December 1970 launch schedule. Lockheed's overtime authorization had been revoked a week earlier. King believed that if the various camera and film tracking problems encountered at Sunnyvale and Danbury were promptly solved, a March 1971 first flight might still be possible. Not until he heard King's opinion did Dr. McLucas officially advise the United States Intelligence Board that the Hexagon schedule had come thoroughly unstuck.[49]

Most of the problems proved to be electronic rather than mechanical or optical, which may have made long-term prospects seem brighter, but that did not lessen the immediate gravity of the situation. Late in September, King named select teams of specialists to review the status of sensor subsystem work and once their preliminary findings had been received sent off additional "tiger teams" to look into the state of affairs at the space vehicle and recovery vehicle plants. Their reports reinforced King's preliminary judgment: if "…no additional significant problems occur…" the first flight midsection should be ready for mating by mid-October and first launch should follow in March. Four months had been allocated for systems integration and checkout at Sunnyvale and Vandenberg.[50]

Although nothing resembling the major testing failures of July and August marred the Hexagon development program for the remainder of 1970, by January 1971 it had become apparent that "March 1971" (which had widely been interpreted to mean "about 1 March") had better be restated as April, and 9 April became the new official target date—although in private session the Executive Committee received advice from Dr. Naka that "about May 10, 1971" was a better estimate. Somewhat less inclined than in the past to accept schedule assurances at their face value, the NRP Executive Committee endorsed Dr. McLucas' action in providing additional insurance against extended Hexagon troubles by authorizing work on a Gambit modification (Higherboy) that would permit Gambit to operate as a makeshift search system by flying at altitudes of about 525 nautical miles. At that distance, Gambit swath widths would approximate those of Corona, and resolution would be about the same. The first of three Higherboy kits ordered for insurance would be ready by November 1971 but would not be needed before April 1972, in the worst possible case. Considered as no more than Corona equivalents, they would add approximately six months to the existing overlap between Corona and Hexagon. They represented, at best, a means of offsetting the consequences of a temporary loss of search satellite capability through an extended delay in Hexagon availability. Higherboy was an expensive but expedient means for providing Corona-scope search capability, with perhaps somewhat better resolution than Corona, but in no sense could Higherboy be considered a Hexagon replacement.[51]

Dr. Naka's cautious appraisal of the worth of "official" Hexagon launch schedules proved sound almost immediately. By the end of March, problems encountered in acoustic and thermal tests of the first payload-vehicle assembly caused program managers to reschedule the initial launch for "not earlier than 3 May 1971," and by April it had become apparent that the four-month allowance for payload integration and checkout should have been seven months. Late in April new delays intervened, and 20 May became the target date. Then on 26 April the program office learned that extended testing of the shutter assembly on the second and third camera payload sections had disclosed that failure was liable to occur after only 28,000 cycles of shutter operation. Colonel Buzard sadly advised Brigadier General Lew Allen, new Program A director, that because the shutters in the payload then being prepared for launch had experienced 20,000 and 28,000 cyclic operations respectively, there was a high probability of shutter failure on orbit. The design, he said flatly, was marginal. He therefore proposed to delay the first launch until at least June.

Allen reacted immediately. Categorizing the possibility of on-orbit failure as "unacceptable," he halted launch preparations. Perkin-Elmer estimated that three weeks would be required to modify and retest the shutters.

The problem, when diagnosed, was almost simplistic. The shutters were of focal-plane types, with the opening and closing blades operating in separate slots in a rail and overlapping at the end of their travel. The shutter blades were .010 inches thick, and the slots .015 inches wide. Bearing surfaces on the shoes on the closing blade were .015 thick. There simply was insufficient room for both blades and bearing shoes when the blades overlapped at the end of shutter travel. The correction was simplicity itself: remove the end of the blade so that it no longer rubbed on the rail; blade failure (from that cause) thus became impossible. Accelerated tests of the modified blade assembly proved it capable of surviving 110,000 cycles of operations. But diagnosis and shutter modification (and retesting) had chewed up so much time that "about 14 June" had to become the new launch target date. (Because Hexagon payload vehicles could not be trucked over California highways on weekends, when traffic was heaviest, and because the payload would not be ready for trucking before 28 May, four additional days delay were imposed by the unfortunate coincidence of the Memorial Day weekend and the completion of payload testing at Sunnyvale.)[52]

But that was the last. Payload delivery was on schedule, pre-launch checkout was almost uneventful, and on 15 June 1971 the first Hexagon satellite went into orbit. Carrying Hexagon from program approval to first launch had taken five years rather than two and had cost rather more than twice as much as initially estimated, mostly for camera development, which cost three times as much as the CIA had anticipated,[53] but a launch had been brought off. And in the end the critical scheduling estimates provided by Dr. Robert Naka and his associates in 1969 had proved remarkably accurate: Hexagon did indeed fly in June 1971 (the "95-percent probability" date), and it did indeed function successfully (the "75-percent confidence" evaluation).

HEXAGON: INITIAL OPERATIONS

Although first flight did not occur in October 1970, as anticipated, Hexagon operations, when they began, conformed in other respects to careful plans designed to meet that deadline. Operation of Hexagon would be as complex as the management and hardware and software problems that had proved so troublesome in the months between April 1966 program approval and June 1971 first flight. The functional and organizational interrelationships of Hexagon operations would have astonished reconnaissance program managers of the early 1960's, when verbal agreements and informal memoranda constituted the bulk of operational program documentation.

The list of organizations participating in Hexagon operations was awesome—even if only principals were counted. It included COMIREX* (the United States Intelligence Board—USIB—Committee on Imagery Requirements and Exploitations); Eastman Kodak; the Global Weather Center (GWC) of the Air Force Air Weather Service; the Imagery Collection Requirements Subcommittee (ICRS) of COMIREX; the National Photographic Interpretation Center (NPIC); the Office of Special Projects (OSP), CIA, the Satellite Operations Center (SOC) of the National Reconnaissance Office (NRO); the Satellite Test Center (STC) of the Air Force Satellite Control Facility (SCF); the Sensor Subsystem Project Office (SSPO) of CIA's OSP; the System Program Office (SPO) at the NRO's Directorate of Special Projects (SAFSP); the Air Force Special Projects Production Facility (SPPF); and the U.S. Army Topographic Command (TOPOCOM). The acronyms alone were enough to engage the attention of a trained philologist.

Both the System Program Director (General Allen at the time of first launch—Colonel Buzard was Program Manager) and the CIA's Director of Reconnaissance (John Crowley) reported to Dr. Flax for purposes of managing the operational aspects of Hexagon. The System Program Office (Los Angeles) and the Sensor Subsystem Project Office (Langley, Virginia) were respectively responsible for mission operational software (computer programs) and participation in the development and analysis of the software. CIA's OSP developed simulation and special study software, and the Satellite Operations Center participated in various aspects of software development as required.

Mission guidance for operational activities came from ICRS/COMIREX, which also provided any details of intelligence requirements not defined previously by USIB standing requirements. The Satellite Operations Center selected launch dates, orbits, and mission objectives to satisfy general intelligence requirements or such special mission requirements as might from time to time be levied. The Office of Special Projects (CIA) provided pre-mission software, and the System Program Office determined vehicle performance characteristics, established flight objectives, defined operational constraints, and provided for vehicle launch preparation and mission operations preparations.

The Satellite Test Center, in support of the System Program Office, constructed mission profiles and a software data base and performed mission software rehearsals. NPIC furnished target lists. The System Program Director exercised complete responsibility for Hexagon operations from launch through recovery. Assisted by the Sensor Subsystem Project Office, the Hexagon Operations Command Post (part of the Satellite Test Center) conducted on-orbit operations.

Global Weather Center was responsible for providing weather forecasts for each one-eighth of the full Hexagon swath width of photography for each orbital revolution, expressing the forecasts in terms of percentage probabilities that any area would be more or less than 90-percent cloud-free.

Eastman Kodak and the Special Projects Production Facility (a superbly equipped photographic processing laboratory at Westover, Massachusetts) each processed two working prints of each set of negatives. The National Photographic Interpretation Center provided preliminary readouts of the film returned by the first and third operational reentry vehicles; Eastman Kodak and the Special Projects Production Facility (with NPIC participating) did the actual film processing and distribution. TOPOCOM provided the final operational report on cloud cover during flight, the Satellite Operations Center evaluated mission accomplishments, and the System Program Director furnished a post-flight analysis of operations for all but the camera system, which was the analytical responsibility of the Sensor Subsystem Project Office.

---

* (Acronyms and organizational abbreviations generally have not been used on these pages, except for such often-used sets of initials as NRO, NRP, USAF, and CIA. The following brief summary of operational program participants and their responsibilities is so dominated by organizations known almost exclusively by their abbreviations that it is not feasible to continue that felicitous practice, however desirable. Some acronyms are so well entrenched in conversational usage in the intelligence community that even constant users have to stop and rummage through their memories when asked to provide the full titles of such as COMIREX, SPPF, and ICRS. The reader baffled or infuriated by bureaucratic fondness for acronyms and their verbalization may pass by this section without appreciably weakening his understanding of the Hexagon program. The section has been included in deference to the canons of historiography: some muddled scholar may someday need to know what element of jargonese referred to what organization. R. P.)

Software capabilities resident in Hexagon included three simulation programs called CRYSPER, HAMPER and HSIM, relating respectively to sensor subsystem performance, mission performance, and camera operations. TUNITY was the name assigned to the mission targeting, command and control, and reporting program; trans-operational mission reports were mostly based on TUNITY products. The probability that mission objectives were being satisfied was calculated through use of a program called ACCOMP.[54]

Preparations for the first Hexagon launch had not gone unremarked by the press, which was scarcely surprising if only because the Titan IIID launch vehicle was so enormous (although it used the same booster core that put Gambit-3 in orbit). Oddly enough, none of the major newspapers of the country noticed the operation. Aviation Week printed a small post-launch notice that singled out Hexagon as a previously untried system but completely misstated mission and function. A feature writer for the San Jose News represented Hexagon to be "a giant super spy satellite known as 'orbiting Pueblo'..." and alerted local residents to watch for "the most volcanic blastoff ever witnessed on the West Coast." The imaginative writer attributed to the satellite the combined capability of being able to "photograph the whiskers on the chin of a Soviet general..." and "monitor whispered conversation from 115 miles in the sky"—which might have been true for some Soviet general with a waist-length beard who was sunning himself on a well-lighted black-sand beach just as Hexagon passed directly overhead, cameras operating, but was in other respects somewhat exaggerated.[55] The Air Force routinely announced the successful launch of a "Department of Defense satellite" after Hexagon went into orbit and a few newspapers printed the now-customary paragraph reporting that event, but nothing in the way of real publicity disturbed the launch.[56]

Perhaps all of the trauma and disorder fated for Hexagon had been used up in the exhausting gestation and development phases; perhaps by 1971 reconnaissance satellite development had become more a science than an art. But in any case, the launch and orbital operations of Hexagon 1201 were as nearly flawless as any first launch of the decade. Camera operations presented "only minor problems," and until the final minutes of film capsule recovery there was nothing in the mission approaching drama. The C-130 retrieving aircraft nearest the first descending capsule easily spotted the target but the pilot elected to let it fall into the ocean after observing that the parachute was badly torn and descent rate was very rapid. A helicopter crew retrieved 1201-1* from the sea less than 30 minutes later, intact and undamaged.

As with other untried systems, the first few Hexagon flights were primarily intended to provide data on capabilities and operational problems. In that sense, useful photography was a bonus. But unlike Gambit, the last preceding major photographic satellite system to enter the service of the National Reconnaissance Program, initial Hexagon operations were also planned to return as much overflight photography as possible. Much of the film captured images of ground targets distributed over parts of the western United States. But Hexagon carried more than 200,000 feet of 6.6-inch film, and even if the cameras had operated randomly it would have been difficult to expend 50,000 feet of film (the quantity contained in each recovery capsule) without managing to photograph some targets of interest to the intelligence community. In the case of Hexagon, of course, exposure was never random in character; camera operations were precisely calculated to provide photographs of denied areas. Dr. McLucas had his first look at the product on 22 June. He immediately advised all Hexagon program participants that it was outstanding, representative of a great technical achievement, and of remarkable value.[57]

As compared to other reconnaissance satellite first flights, Hexagon 1201 may have been relatively trouble-free, but there were operational difficulties nonetheless. Battery overheating, apparently the product of solid-rocket exhaust contamination of reflective surfaces, perturbed on-orbit control to some extent, and the parachute malfunction on 1201-1 was symptomatic of a potentially serious problem. Capsule 1201-2 also developed parachute problems, although in that instance an air catch (26 June) proved feasible nonetheless. But 1201-3, which reentered on 10 July, was another matter. All went well to the instant of main parachute deployment, but at that point a catastrophic overload developed, the parachute lines failed, and the capsule hit the ocean with such great force that the impact ruptured flotation devices. Before nearby helicopters could arrive, the capsule sank to the ocean floor some three miles below.

The battery overheating problem foiled plans to extend the first Hexagon mission to 45 days. By early July, degradation of the primary batteries threatened a shift to reserve batteries (carried on the first mission

---

* In order to limit confusion in discussions of the four-capsule Hexagon system, it seems advisable to adopt here the convention used in Hexagon mission reporting, identifying the mission by number (in the 1200 series, starting with 1201) and the mission phase and reentry vehicle by sequence of capsule use. Thus "1201-1" identifies both photography returned by the first of the four capsules to be retrieved in the first Hexagon operation and the capsule itself.

only) and there were indications that the batteries used to ignite reentry pyrotechnics were failing. When one set of pyro batteries did fail, on 14 July, Buzard had the cameras operated at every possible opportunity for the next eight hours and then recovered 1201-4 on Friday, 15 July, after Hexagon had been 31 days on orbit.

The battery problem had constrained camera operations earlier in the mission, indications of overheating being responsible for a decision to limit photography to about 15 minutes for each of four orbits, a provision that was subsequently relaxed and eventually cancelled. Nevertheless, the availability of reserve batteries protected against a total failure of Hexagon 1201 if the main batteries and solar panels were to prove defective. Hexagon 1201 had been programmed for only 30 days of orbital operations, and realization of a 31-day mission represented performance marginally better than planned.

During the transfer of film take-up operations from 1201-3 to 1201-4, some undiagnosed disorder in the film transport mechanism caused two brief automatic shutdown operations, but resort to ordinary recovery measures restored the cameras to full function shortly after each incident.

On balance, Hexagon mission 1201 had to be adjudged an outstanding success. Returned photography from 1201-1 alone contained coverage of more than two-thirds of all known Soviet missile sites and one set of photography taken in one pass over Albania was sufficient to permit identification, by class and type of weaponry, of that country's entire inventory of aircraft and ships. The battery overheating defect seemed likely to be cured by battery repositioning. Film transport difficulties appeared to be of no great consequence, although when the recovered capsules were unloaded it began to appear that less film had been exposed than planned, the outcome of faulty transport mechanism operations that caused film to twist and double over itself from time to time on both sides of the platen.* But with 50,000 feet of film available for each of the four mission phases, the occasional loss of a hundred feet here and there did not seem important for a first mission.

Parachute problems were quite another matter; only one of the four recoveries (1201-4) had been free of parachute malfunctions, two of the capsules had gone into the ocean (1201-1 and 1201-3), and one had been lost altogether (1201-3). Happily, the damaged parachutes of 1201-1 and 1201-2 had both been retrieved, as had the ablative cone of 1201-2, so analysis of the problem did not have to be conducted on the strength of telemetry and photography alone.[58]

Analysis of parachute defects, and later redesign and test activities, continued for several months after 1201-4 had been retrieved. Modified parachutes were eventually provided for 1202, and a new canopy and rigging design was adopted for 1203 and later Hexagons.** The failures, it appeared, had been the products of design and testing oversight. Because tests of the parachute assembly used for 1201 had not fully explored the high-shock region of initial parachute deployment, the parachutes used for 1201 were at best marginal. That three of the four capsules had been retrieved later began to seem almost miraculous; by all odds, the ratio should have been reversed. The main parachute lines had been overstressed by a factor of about two. Discovery of that situation, and the provision of adequate parachutes, eventually contributed to a decision to delay launch of Hexagon 1202, although in fact 1201 had returned so much still undigested information that in July the head of the Defense Intelligence Agency suggested to Dr. McLucas that Corona and Gambit operations would more than satisfy intelligence needs for the moment.[59]

And Hexagon 1201 provided an opportunity no earlier reconnaissance satellite could have matched: for the first time the United States seriously attempted to retrieve a space capsule from the bottom of the Pacific Ocean.

The capsule at issue was 1201-3, which lay at a depth of about 16,000 feet on the floor of the Pacific Ocean off the Hawaiian Islands. Capsule designers were confident that the water impact had not shattered the capsule, although it was quite likely to have been damaged and would not be water-tight.

The feasibility of recovering 1201-3 was first considered almost casually in a 27 July conversation between Dr. Naka and Carl Duckett of the CIA. Intrigued, Duckett discretely asked the Navy if the deep-submersible Trieste II, an experimental submarine of considerable versatility, could do the retrieval task. The Navy assured Duckett that the Trieste II could operate safely to depths of 20,000 feet, could quite probably manage a "hook and cable" retrieval operation, and that the precise location of the capsule could probably be established by Scripps Institute undersea survey ships then under charter to

---

* Some unexposed film was programmed: film rewind between camera operating phases was not scheduled for Hexagon 1201 in the interests of simplifying first-flight operating modes. Film twisting and overlapping caused the metering instruments to register more film on the take-up spools than actually reached them.

** The redesigned parachutes were originally planned for incorporation in 1205, but schedule slippages caused by other factors eventually allowed their addition to 1203.

the Navy—at a cost of only $100,000. The Trieste II would be provided cost-free, if wanted.

Upon hearing the first informal suggestion that the capsule might be recovered, an NRO staff officer *asked Eastman Kodak if the film conceivably could be exploited after retrieval. The answer, surprisingly, was "yes"; the edges of the tightly rolled film would swell upon exposure to sea water, thus sealing off the center of the roll. In Eastman's judgment, a "considerable portion" of the film might survive. Because the weather over the Soviet Union had been good while the film returned by 1201-3 was being exposed, and because the film included coverage of some regions of particular interest to the intelligence community, its recovery might be highly worthwhile. D. W. Patterson, the CIA's program director for Hexagon sensor subsystems, had earlier obtained EK's assurance that the film could be safely despooled by hand. He concluded that if a search could be started by late August there would be a "good chance" of recovering useful film from 1201-3 during September.[60]

Once the informalities had been disposed of, Dr. McLucas formally asked R. A. Frosch, Assistant Secretary of the Navy (R&D), to authorize use of the Trieste II and Scripps Institute survey ships in the recovery effort. Frosch assured McLucas that the Navy would be "pleased to assist" and that the exercise would cost the NRO very little.

The effort, once begun, proved to be somewhat more troublesome than first assumed. An initial afterthought prompted a decision to design and fabricate a special basket container with claw-like clamps rather than to make retrieval dependent on hooks, cables, and the eyebolts of the capsule. Because weakening or disintegration of the magnesium parts of the capsule would occur during exposure to seawater, an enclosing "basket" would provide greater assurance that the capsule could be brought to the surface more or less intact. Delays imposed by "basket" procurement, bringing the Trieste II to the scene, and precisely locating the capsule delayed the start of recovery operations until winter weather arrived, in December 1911. (A start had been scheduled for late October.) But thoughts of abandoning the attempt could not realistically be entertained once it had begun. As Colonel Buzard pointed out, Soviet interest had almost surely been piqued by the unconcealed activities of such sea-bottom survey ships as the White Sands, Apache, and De Steiger, and given that the Soviets had precisely the same rights in the open Pacific as the United States, a Soviet effort to recover whatever the U.S. had been seeking was not at all inconceivable.

* U.S. Army, assigned to the NRO.

The Soviets were known to be able to operate deep-submergence vessels at depths as great as 33,000 feet and were notoriously persistent. (A Soviet ship had then been keeping station over a sunken November-class submarine in the North Atlantic for more than 18 months, presumably to preclude any U.S. effort to recover hardware.) In such circumstances, Buzard argued it would be foolhardy to abandon the effort to retrieve 1201-3. Should that be unavoidable, however, he suggested that the on-station ships should pretend that the recovery operation had been successful—sending congratulatory messages, tying off logistics arrangements, and otherwise presenting a bold front.

Dr. McLucas decided to persist. With the passing of bad weather in the Central Pacific in the Spring of 1972, the Trieste II returned to the scene and recommenced its deep-sea search. The survey ships reported success in locating what appeared to be capsule 1201-3 in mid-April. During the afternoon of 26 April, the Trieste II completed a two-hour submersion operation and after a search of three and one-half hours sighted first debris and then the actual capsule at a depth of 16,400 feet. Three and one-half hours of careful maneuvering preceded basket closure and the start of a cautious ascent. More than nine hours after starting its dive the Trieste II surfaced. Unhappily the action of surface waves proved too much for the now-fragile capsule structure, which broke into pieces so small that most fell through the tines of the recovery device. Only some inconsequential bits and pieces remained.

Disappointing as the outcome may have been, it was one instance in which the death of the subject following a successful operation proved a thoroughly acceptable alternative to recovery. Nothing suggestive of U.S. reconnaissance capability remained for others to find. The eight-month effort had to be considered at least a partial success even if deterioration of the capsule had prevented full recovery: as McLucas told Frosch, the Navy had established and demonstrated "a unique capability vital to the security of the United States" that might conceivably be called into use again if the circumstances so warranted.[61]

Hexagon 1202 had initially been scheduled for launch about three months following reentry of the final capsule from 1201, but engineering modifications dictated by the performance of the first Hexagon payload were expensive of time, and the alternative of launching one of the few remaining Corona vehicles seemed preferable to a hasty patch job. In the event, assurance of a successful mission would have been lessened if battery overheating problems and the parachute malfunctions encountered during operation of Hexagon 1201 went uncorrected, and both required more effort and took longer than had initially been planned. Nor was the urgency of Hexagon coverage as pressing as had been anticipated; even though one capsule of film had been lost, the amount of film recovered from the first operation inundated photo interpreters. The notion of providing continuous coverage of the Soviet Union and Mainland China by keeping either a Gambit or a Hexagon in orbit at all times had to defer to the realities of staff and dollar resources; processing and evaluating Hexagon and Gambit film in the quantities that the two systems were capable of returning would force enlargement of the cadre of photo interpreters, a course neither the NRO budget nor NPIC training facilities could accommodate. Moreover, the concept of continuous coverage implied a capability for crisis reconnaissance rather than constant operation of orbiting reconnaissance vehicles, and systems other than Hexagon appeared to be better prospects for that assignment. (By 1971 the premise of near-real-time readout by means of an electro-optical imaging system had proceeded to a system development phase, 1976 operational availability having been approved as a schedule goal.)

What with lessened pressure for early launch of Hexagon 1202, some difficulties of system modification, and the availability of one additional Corona system for use in an emergency, it was December 1971 before Hexagon 1202 reached the launch stand. Preparations for a 21 December launch were aborted by elusive booster-system electrical problems, and correction was so lengthy that a complete revalidation of the spacecraft eventually had to be undertaken. The resulting delays caused program managers to reschedule Hexagon 1202 for a 19 January launch, a date that had to be slipped by one further day when last-minute checkout operations uncovered a minor system fault. Hexagon 1202 went into orbit on 20 January 1972 without encountering any major problems. There proved to be no warrant for qualms about the complete reliability of the modified parachute recovery system; 1202-1 and 1202-2 reentered and were uneventfully retrieved on 26 January and 8 February respectively, just as planned. But attitude control subsystem effectiveness had begun to degenerate by early February, so flight managers reluctantly reprogrammed Hexagon 1202 for 39 rather than 45 days of operation. (Premature control gas exhaustion owing to frequent vehicle repositioning maneuvers was assumed to be the source of the difficulty.) A much more serious problem occurred immediately following the start of camera operations for 1202-3; the film being fed through the aft camera developed a ragged tear that quickly became a film break, and for the remainder of the mission only the forward camera was operable.

Capsules 1202-3 and 1202-4 reentered routinely and were recovered without further incident on 17 and 28 February respectively, but imagery was entirely monoscopic. (Worry that the reentry vehicles might be unstable because one of the two take-up spools of each was empty proved unfounded.) De-orbiting of 1202-4 eventually became dependent on the back-up recovery system ("Lifeboat") with the final exhaustion of attitude control gas on the last day of the operation, but the reserve system functioned with commendable effectiveness and no unexpected problems developed.

On the whole, the film imagery returned by Hexagon 1202 was somewhat better in technical quality than that of 1201, displaying a best resolution and a "normal" resolution ranging between 30 and 33 inches. Because of poorer weather and sun angle, it contained no better detail in ground coverage, however. A variety of minor defects marred the operation, some of them the apparent consequence of having camera subsystem remain inactive on the launch stand some four weeks longer than planned, but on balance 1202 had to be counted a successful operation. The principal qualifier in that judgment was the major camera system malfunction midway through the mission, the product of film breakage. In terms of film lost or unused, 1202 and 1201 were about equal.

Diagnosis of the cause of the film failure was difficult. There appeared to be no reason to conclude that it was related to rewind operations first attempted during mission 1202. (In the interests of mission success on 1201, film had been fed through to the take-up cassettes continuously, rewind operation being bypassed. Although that operation had caused wastage, film being passed through the system while the camera was inoperative, it obviated concern for the proper functioning of the rewind mechanisms at a rate of 55 inches per second, probably the most complex elements of the Hexagon's camera system.) Inspection of recovered film suggested that some large particle of foreign matter had become enmeshed in the film transport mechanism of the aft camera, causing a puncture that quickly became a tear when tension increased during rewind and forward transport of unexposed film immediately following the start of 1202-3 events. But the diagnosis had to be tentative because there was evidence of several malfunctions in film transport. Relatively large quantities of film were twisted, overlapped, and tangled on the take-up spools. (Some sections of recovered film had to be torn loose from the spindles during despooling, being so tightly jammed between the spool hub and the spool framework that they defied ordinary removal efforts.) Part of the film damage apparently resulted from an unprogrammed spool rotation after film take-up had been transferred from one recovery capsule to the next in line.

In addition to the film break that had to be counted as the principal defect of Hexagon mission 1202, analysts cited two other major problems. One was a repetition of the thermal heating anomaly that had affected the batteries during mission 1201. Although the batteries of 1202 had been relocated to avoid heat imbalance caused by launch debris that contaminated the reflectant paint, they proved to be almost as susceptible to overheating as those of 1201. Careful control of the angle of exposure (beta angle) of the battery section to solar radiation kept any major difficulty from developing in 1202, but that requirement imposed unwanted constraints on the launch window for the Hexagon vehicle and contributed to the premature exhaustion of attitude-control gas.

The second problem was attitude control. Post-mission analysis suggested that the failure of the reaction control system late in the mission, forcing early recovery of 1201-4, had probably been caused by contaminated hydrazene. The hydrazene at Vandenberg proved, upon inspection, to be "less pure than expected." The most immediate way of correcting for the problem would be to lessen demands on the reaction control system during mission 1203 by flying at the 100-mile altitude of 1201 rather than the 82- to 85-mile altitude of mission 1202. That would somewhat adversely affect resolution potential, but it would reduce the requirements for vehicle maneuvering and improve the potential of flying a full 45-day mission.

The post-mission critique on Hexagon 1202 highlighted another problem, but one unrelated to system functioning. Midway through the operation, intelligence officials in Washington had cancelled pre-mission requirements for operations that would have completed the required annual survey of selected Soviet land areas. The change had been justified in terms of potential savings in film, given the availability of generally adequate earlier coverage of several sensitive areas, but in fact Hexagon operations were in no way constrained by film supply during 30- to 45-day missions. Program officials suggested, rather bluntly, that it was not the function of the intelligence requirements community to manage mission operations, and that it was particularly inappropriate for the Committee on Imagery Requirements and Exploitation to intervene in ongoing operations that were so heavily dependent on pre-programming. [62]

Hexagon 1203 did not go into orbit until July 1972, more than four months after the final bucket from 1202 was retrieved. The search-mission gap created by the delay was partly filled by launch of the last Corona in May and Hexagon program managers took advantage of the respite to incorporate in 1203 several system modifications that had originally been planned for later vehicles. The spacecraft and sensor system of 1203 had encountered more than the ordinary run of qualification difficulty during final tests late in 1971, but those problems were not direct contributors to the launch delay. Test indications of faulty operation in various aspects of film transport did portend similar problems on orbit, but the problems seemed to be almost basic to the complex film transport mode adopted for Hexagon. System validation tests applied to 1203 were somewhat more carefully attuned to film transport and attitude control functions than had been the case for earlier Hexagon systems, but that was no more than ordinary prudence given the difficulties actually experienced in those earlier launches.

The parachute redesign undertaken following the unhappy outcome of Hexagon 1201 capsule reentry operations reached fruition in time to permit 1203 to take advantage of it. Delays in the readiness of 1202 and 1203 permitted the incorporation of redesigned parachute systems in the recovery capsules of 1203 rather than 1205, as had been initially planned. The third Hexagon was also the first of its kind. Problems with platen positioning and film tracking slowed final checkout of 1203, causing the launch to be put off from June to July. The space vehicle finally qualified for launch and left the factory for Vandenberg on 21 June. No serious checkout difficulties occurred thereafter, and 1203 went routinely into orbit on 7 July 1972.

Reentry vehicle 1203-1 deboosted and was recovered by air catch on 15 July after storing film from the first eight days of operation. During the next 14 days, until 1203-2 was recovered, the reaction control system experienced excessive attitude-control gas consumption. The problem became acute while film was being exposed for return in 1203-3; flight controllers eventually had to switch to the backup attitude control system. Concurrently an emergency shutdown occurred in the aft camera system, causing curtailment of stereo photography. Recalling the catastrophic failure that had marred Hexagon 1202, the program office decided to satisfy as much of the coverage requirement as possibly by using monoscopic photography.

The wisdom of that decision became obvious upon inspection of capsule 1203-3 following its 12 August retrieval. Severe film folding was the determinate cause of the stoppage in aft camera operations. The source of the problem appeared to be misalignment of film on the transport rollers. Passage of a section of folded film past the metering capstan had distorted measurements of the lengths of film being transported, causing the control mechanism to call for SLOW speed take-up in combination with fast film feed.

The looper system (which held excess film passing through the camera section) promptly overfilled and an emergency stoppage resulted. The remedy, employed for the balance of mission 1203, was to operate the film transport system at slow speeds. But the prescription had a price: unless rewind speeds reached 55 inches per second, unexposed film passed to the take-up reels. Yet with the system operating at full speed, intermittent accordion folds occurred in the film, each more than 50 feet long. Even at slow speeds the transport system continued to double film back upon itself, but the folds averaged only about 3.5 feet in length. Before the emergency shutdown, roller misalignment had caused edge folds that in one instance extended for 1800 feet and in another affected 400 feet of film. Although it was not possible to determine precisely what sort of misalignment had caused the problem, the malfunction seemed to have occurred in the last set of cluster rollers over which the film passed before entering the take-up spool of 1203-3.

Another emergency shutdown occurred while film was being fed into 1203-4; it was cleared without great difficulty and stereo camera operations continued to the end of the mission. But hopes for a successful 65-day mission were dashed when the backup attitude control system began to use propellant at an abnormally high rate. Discretion-minded flight controllers brought 1203-4 down on 2 September, eight days sooner than programmed, but still 57 days after mission start. After film recovery had been completed, the satellite operations group began a series of attitude control experiments using the capsule-less orbiting vehicle, attempting to find the source of the control gas wastage. They were able to maintain control for another 12 days. Evaluation of telemetered data indicated that valve seats in the control gas system had somehow become so thoroughly fouled that leakage was continual.[63]

Hexagon 1204 was like its immediate predecessor in many respects, although film tracking problems were fewer. Following a 10 October 1972 launch, operations during the first phase of the mission were quite routine. Reentry vehicle 1204-1 was recovered on 21 October without incident. Early in the second phase of the flight, telemetry indicated an incipient failure of attitude control forcing temporary reliance on the backup system. A film tracking problem caused an emergency camera stoppage on 8 November, three days after 1204-2 had separated and reentered, but careful manipulation of control devices limited the shutdown to a single day and once the jam was cleared the problem vanished. Until that time the camera system had been operated in the slow-rewind mode and the vehicle under maneuvering restrictions in an effort to avoid any recurrence of the film tracking difficulties that had troubled 1203. After 9 November, both constraints were cancelled and the system operated at design film transport speeds and without pointing restraints. Excessive yaw and roll rates were recorded intermittently after 14 November, forcing another camera shutdown and another reversion to the redundant attitude control system. Capsule 1204-3 reentered on 23 November, again without incident. (The new parachute system was functioning magnificently.)

The final phase of mission 1204 proceeded without encountering major problems, although the command system gave cause for some concern on 27 November when it ignored a series of reprogramming instructions. Flight controllers disabled the offending circuitry and proceeded to the end of the flight without further difficulty. Late in the mission (after 9 December), the satellite control group began to inject payload command directions daily rather than on alternate days, as had previously been the rule. The greater frequency of command instructions permitted flight controllers to better utilize weather data in directing camera operations, thus significantly improving the quality of ground imagery. (Hexagon pictures had rather mysteriously been more degraded by cloud cover than earlier Corona pictures. That was in part the consequence of nothing more alarming than a run of bad luck in weather prediction, but it also reflected the fact that Hexagon photographic swath widths were far wider than Corona swaths and thus recorded more clouds as well as more ground area.)

The final capsule of Hexagon 1204 reentered and was recovered on 17 December; primary mission time was 68 days, three days longer than the earlier "extended" goal. Photography was superb, characterized as the best the system could hope to produce.[64]

The fifth Hexagon mission, 1205, was distinguished by the inclusion, for the first time, of the 12-inch focal length mapping camera and a small fifth reentry vehicle to return its exposed film. Originally scheduled to be flown in the 1207, later in 1206, and finally in 1205, the mapping camera benefitted both from faster than expected progress and from the slower than expected rate of Hexagon launches. Nonetheless, the first mapping camera intended for flight failed during thermal vacuum testing early in 1972 and required complete overhaul before retesting. Shutter malfunctions also interrupted qualification testing later that year.

Hexagon 1205 had additional problems with the attitude reference module, telemetry equipment, and other specialized modules of the satellite vehicle. Delays in delivery of a flight-qualified attitude reference module provided the pad of time needed to install the mapping camera in 1205. There was no longer great pressure for an earlier launch date, the returns from the first four Hexagon missions having been sufficiently impressive to support a decision that only three such vehicles need be operated each year. (Budget pressures also contributed to the three-per-year decision.)

Thermal vacuum tests were completed on the entire satellite except for the attitude reference module by 27 November, and acoustic chamber tests by 11 December 1972. By the end of the year only the solar arrays and the attitude reference module remained to be added for flight readiness. After the attitude reference module was finally delivered, the mated satellite was shipped to Vandenberg on 21 February 1973 for a 9 March launch. As delivered, 1205 included relatively large elements of equipment originally planned for initial use on 1206, the launch schedule relaxation having provided time needed to move improved items forward in the program. (Until late January, a 15 February launch date had been scheduled.)

The mission began on schedule with a routine launch and orbital injection. A loss of telemetry data on camera temperature and pneumatic gas caused initial search camera operations to be postponed for a full day past the fifth revolution—the usual starting point. In order to minimize the potential for film transport malfunctions, modest rewind speed and scan angle constraints were maintained. The mapping camera was slated to begin operations on the eighth revolution, but its lens-cover door jammed momentarily. On the next and all but the last few orbits of 1205 the door functioned correctly, but mapping operations for the balance of the mission were restricted to targets below 50° North latitude on the descending portion of each revolution to protect against a recurrence of the temperature problem that had caused the pneumatic door activator to stick.

As in previous flights, the first mission segment was completed without major incident. Capsule 1205-1 was recovered on II March. Shortly thereafter, the propellant leak difficulty experienced in earlier flights became troublesome. Recovery vehicle 1205-2 was de-orbited and successfully recovered on 4 April and flight phase three began before it was necessary to switch to the redundant reaction control system, however. A yaw rate bias developed subsequent to the shift to a backup attitude control system, but mission controllers were able to use image motion compensation to overcome the smear problem thus created. The third and fourth reentry vehicles were successfully recovered on 18 April and 11 May 1973, respectively. The mapping film recovery capsule, carried in a separate compartment in the forward part of the satellite vehicle reentered independently on 20 April.

A number of minor problems occurred with the mapping camera during the course of the flight additional to the sticking camera door. On four separate occasions, shutdown commands were ignored. Eventually flight controllers had to call on an alternative command system to reactivate the cameras. (Redesign of the door actuating circuitry was undertaken immediately after Hexagon 1205 completed its mission.)

About 95 percent of the mapping camera film had been successfully exposed and resolution was some 30 to 40 percent better than had been predicted. The film product of the main cameras was again of excellent quality, approaching in "best" resolution. It seemed even to surpass the product of Hexagon 1204 in one instance (1205-1), but that was the consequence of excellent weather and lighting rather than any optical superiority.[65]

HEXAGON: PROGRAM REORGANIZATION AND PRODUCT IMPROVEMENT (1971-1973)

The success of Hexagon in satisfying the requirements against which the system had been developed could best be judged from the fact that as many as six missions per year had been planned while the system was in evolution, but by 1972 returns from early missions were so satisfying that launches at four-month intervals (three each year) served needs. As early as November 1971 it was apparent that Hexagon and Gambit in combination would return twice or three times as much information as the United States Intelligence Board had formally required, and both systems were susceptible of relatively modest improvements that would substantially extend their on-orbit operational lives. Hexagon, intended for 45-day operational mission, had early demonstrated 60-day capability and there were no obvious technological obstacles to flying 75-day missions. In that event, each Hexagon would perform tasks originally assumed to require 1.8 successful operations.

National Reconnaissance Program managers were not at all displeased by such trends. Hexagon had cost nearly, three times as much money and twice as much time to develop as anticipated when the program was approved, and the cost of operational systems was nearly twice that planned. But the real cost of satisfying requirements for which Hexagon had been specified would approximate early predictions once 75-day missions were achievable, and few outcomes could be more satisfying. Real budget reductions could be enacted. However, it was likely that 75-day missions would overload the interpretation capability of the National Reconnaissance Program, driving costs upward in another area, a possibility that prompted David Packard, Deputy Secretary of Defense, to urge a reduction in the frequency of operating other intelligence collecting systems in order to avoid such problems.[66]

Proposals for significant improvements in various aspects of Hexagon performance entered a phase of serious discussion virtually as soon as Hexagon 1201 had completed its operations. Some of the notions first formally considered within the NRO in August 1971 involved improvements earlier proposed but temporarily tabled because of the urgency of starting Hexagon operations before the supply of Corona systems was exhausted and a gap in search coverage developed. The CIA's sensor project office had concluded by the summer of 1971 that a change in Hexagon configuration should be made effective with the nineteenth system (then planned for launch late in 1976). The spectrum of attractive presumably achievable changes extended from a relatively modest extension of mission life to about 90 days using essentially the original camera system (though preferably simplified in details of film transport and electronics) through a major redesign leading to 145- to 180-day missions (which implied a two-per-year Hexagon requirement). Plainly, a 180-day on-orbit capability would involve, as a minimum, increasing the number of reentry capsules and enlarging film capacity. A complete camera redesign could not be excluded from consideration. The broad goal, established by Dr. McLucas, was to provide for competition in sensor subsystem procurement and to reduce the recurring costs of operating Hexagon.[67]

NRO staff officers considered the CIA's proposals to be rather more optimistic than circumstances warranted, but nonetheless the NRO budget was altered to provide for fiscal year 1973 funds to begin work.[68] The objections were not entirely on feasibility grounds, however. No requirement for missions of more than 75 days had been validated, the effect of having a near-real-time readout satellite operational had not been assessed, and as one NRO staff officer tartly put it, "hardware changes should not be funded for the sake of hardware changes..." Technical feasibility was not a valid reason for making major system changes: timeliness, national requirements, and cost-effectiveness considerations had to be counted in the equation.[69]

Although the happy outcome of early Hexagon flights induced many senior officials to assume that a 75-day orbital life for the system was readily achievable, no such premise was valid in late 1971. General Allen cautioned Dr. McLucas that "...ongoing procurement actions... do not at present include preparations for obtaining this extended life capability." Hexagon vehicles through 1212, then on contract, were designed to satisfy 45-day mission requirements and to have 60-day-qualified components. The orbit adjust system originally built into Hexagon was design limited to 45-day operations at normal altitudes, although provisions had been made for extending to 60-day missions "when desired." The 60-day missions achieved in 1973 were made possible by increasing perigee altitude, with some loss in system resolution and with acceptance of a slightly lessened probability of successful mission completion. The absolute limit of Hexagon life, in its original configuration, was 750 camera operating cycles. (That constraint was imposed by the limited supply of pneumatics required for camera functioning.)[70]

Discussions of potential Hexagon improvements continued for more than a year after they first were proposed. By the spring of 1973 they had progressed to the stage of a potential new camera competition, a possibility created in part by Itek's unsolicited submission of a new camera design remarkably similar in many respects to the Itek design that lost to Perkin-Elmer's Fulcrum based proposal in 1966. Brigadier General David Bradburn, who by that time had replaced General Allen as the NRO's Director of Special Projects, urged Dr. McLucas in April to approve and fund a new panoramic camera definition study by Itek. The goal, as Bradburn saw it, should be "...an alternative sensor subsystem with improved performance and simplified design, and at reduced cost," that could be incorporated in the first "Block IV" Hexagon (still the nineteenth production system).

The source of the proposal was a "HEXAGON Panoramic Camera Improvements Study" prepared by Bradburn's staff in the spring of 1973. He proposed to sponsor a technical evaluation that could conceivably lead to a formal Block IV Hexagon competition in May 1974. The new camera would incorporate a faster (f/2.0) 60-inch focal length lens than the Perkin-Elmer design, reduced film velocities at the film plane (the chief problem generator in the original Hexagon), but full compatibility with other principal elements of the existing system (film supply, telemetry, take-up section, test equipment, and vehicle design). Its principal attraction, apart from potentially better resolution arising in improved optics that permitted use of slower high-resolution film, would be to eliminate the troublesome film rewind operation designed into the original Hexagon camera.

After considering the proposal in detail and insuring that its approval would not create major funding problems, Dr. McLucas on 4 May 1973 approved starting the study. (Dr. J. R. Schlesinger, newly-installed CIA director, had informally approved the approach in the course of a 17 April discussion of Hexagon improvement potential.)[71]

A second development of mid-1973 that had considerable significance for the future of Hexagon was the transfer of camera subsystem responsibility from the CIA's Sensor Subsystem Program Office to the NRO's Program A, the West Coast Directorate of Special Projects. Proposals for that shift of authority had been informally considered two years earlier and had reached the stage of a formal plan by October 1971.

The motivation for the transfer was not obscure. On 23 September 1971, President Nixon approved a plan to develop a highly ambitious near-real-time readout reconnaissance satellite,* on a schedule that called for initial operations during 1976. Most was to be a CIA responsibility. With limited resources for managing reconnaissance satellite development, the CIA faced a future that encompassed both the most costly and complex of ongoing reconnaissance satellites (Hexagon) and a yet more costly and complex future system. In the circumstances, handing over Hexagon to the NRO's West Coast establishment seemed a wholly sensible course.

The plan for transferring Hexagon sensor responsibility reached Dr. McLucas on 21 October 1971, the only point of residual disagreement being whether responsibility for Hexagon 1207 through 1212 (Block II) should be reassigned on 1 July 1972 or 1 July 1973, the CIA's principal spokesman holding out for the later date.** There was no controversy about the transfer of responsibility for Block III Hexagons (1213 through 1218) or the still undefined Block IV model; all were agreed that action should be completed as rapidly as possible so that orderly planning for an improved Hexagon might proceed.[72]

Dr. McLucas chose to accept the argument for transitory CIA retention of responsibility for systems 1207 through 1212. On 29 November he assigned immediate responsibility for Hexagon 1213 and later systems to General Allen and expressed to the director of the CIA's reconnaissance programs his wish that arrangements be made for the timely transfer of systems 1207 through 1212. (Only the contracts with Perkin-Elmer were at issue; all other CIA-managed Hexagon contracts were shifted to Allen's custody at once.) Dr. McLucas hoped to complete all actions essential to the reassignment by the summer of 1973, exempting only those functions (like mission simulation and statistical prediction studies) in which the CIA had an unduplicated competence.

The formal transition plan, completed and forwarded for NRO and CIA approval in March 1972, provided very largely what Dr. McLucas had suggested in response to the initial plan the preceding October. Systems 1207 through 1212 would be transferred (to the Director, Program A—the West Coast group) effective 1 July 1973 in accordance with contractual agreements with Perkin-Elmer which were to be formalized no later than 1 September 1972. Certain specialized Hexagon-related activities of the CIA were exempted (in addition to the software work) and the CIA would retain full responsibility for Hexagon systems through 1206,*** but virtually all else would be captured by the shift. The CIA agreed to provide full engineering support to Program A during the transition period. Interestingly, in light of the strong feelings that had existed at the time Hexagon working relationships were first established in 1966, the transfer agreement explicitly provided for "a free exchange of information between CIA/OSP and SAFSP on all elements of the HEXAGON Program to be transferred."[73]

The Program A contract with Perkin-Elmer for systems 1207 through 1212 actually became effective on 1 December 1972 rather than 1 September, as earlier planned, but other aspects of the transfer proceeded very nearly on schedule. The overlap of CIA-Program A efforts was generally smooth and effective. The only substantial change in procedures that resulted from the transition was a shift of acceptance point for the camera systems from the Perkin-Elmer plant at Danbury, Connecticut, to the Lockheed facility at Vandenberg. The original justification for accepting camera systems at the Perkin-Elmer site had been the need for the contractor to deal directly with chamber test problems, part and component failures, and similar events, and the desire on the part of the Hexagon program office to make test qualification rather than extreme schedule urgency the prime

---

* Generally referred to as "the EOI system," for electro-optical imaging.

** Participants in the preparation of the transfer plan were General Allen, Harold Brownman (CIA), Dr. Naka, (NRO Comptroller), and then-Colonel Bradburn (Director of the NRO Staff).

*** In March 1972 the flight schedule called for Hexagon 1206 to be launched in April or May 1973; various technical problems and a major revision of coverage requirements delayed that event past 1 July 1973 and 1205 became the last Hexagon to be launched in fiscal year 1973. Nevertheless, the CIA kept responsibility for 1206.

contract incentive. As Perkin-Elmer became more familiar with space program operations, the need for the special arrangement at Danbury disappeared.*

On Friday, 29 June 1973, L. C. Dirks, the CIA's senior Hexagon-responsible official, advised General Bradburn that effective 1 July all responsibility for the camera systems for Hexagon 1207 through Hexagon 1212 was transferred to his organization. The Agency would continue to monitor the delivery and operation of 1206 (still awaiting launch), but funds transfer would be complete by 6 July 1973, and that would effectively end the CIA role in Hexagon development and operations.

On the following day, General Bradburn notified Dr. McLucas of his formal acceptance of the assignment and sent a final message to Dirks: "I congratulate you on the success of the program under your leadership and I assure you we will do our very best to continue that proud record."[74]

---

* It will be recalled that the CIA arrangement with Perkin-Elmer in 1966 and 1967 was also influenced in some part by the residual distrust of the Program A staff by CIA satellite specialists, a consequence of the factionalism that had marked CIA-NRO relationships in 1964 and 1965.

Endnotes

1. Details of the E-5, E-6, and Corona-Mural programs are to be found in the chapters devoted to those topics.

2. See ltr, B. McMillan, DNRO, to V/Adm W. F. Raborn, Dir CIA, 3 May 65, no subj, in DNRO files.

3. Additional details of relevant Corona, E-5, E-6, M-2, J-3, and Lanyard developments are included in chapters dealing with Corona and Samos programs. The management controversies of 1963-1966 are described in Volume V, this study. See also: Memo, E. M. Purcell, Chm, Recon Panel, to DCI, Jul 63, subj: Panel for Future Satellite Reconnaissance Operations; memo, M/Gen R. E. Greer, Dir/Progm A, to DNRO, 15 Apr 63, subj: Comparison Evaluation, and encl, Report of the Findings of the Ad Hoc Group Appointed to Evaluate Potential Systems for an Improved Search Type Satellite Reconnaissance System, Apr 63; memo, E.G. Fubini, DDR&E, to USecAF, 30 Jun 64, subj: Broad Coverage System; MFR, E. Fubini. "Dictated in Mr. McCone's Presence," 13 Jan 64; memo, C. B. Clifford, Chm, FIAB, to the President, 2 May 64, subj: National Reconnaissance Program; memo, B. McMillan, DNRO, to D/SoD, 12 Jun 64, no subj, all in DNRO files.

4. Memo, A. D. Wheelon, D/Dir S&T, CIA, to DCI, 31 Aug 64, subj: Conduct of the FULCRUM Program; memo. E.G. Fubini, DDR&E, to SAFUS, 3 Jul 64, subj: Broad Coverage System; MFR, B. McMillan, DNRO, 7 Jul 64, subj: CIA Management of Satellite Projects. The SP-AS-63 episode is detailed in Vol IIB, this mss (Ch XI).

5. Ltr, C.R. Vance, D/SOD, to DCI, DNRO, 8 Jul 64, no subj; memo, A. D. Wheelon, DI Dir (S&T), CIA, to DNRO, 9 Jul 64, subj: Funding for Project FULCRUM; memo, Col J. C. Ledford, Dir/Progm B, to DNRO, 10 Jul 64, subj: Addendum to Pgm B's FY 65 Budget.

6. See ltr, McMillan to Raborn, 3 May 65; SOR Description and Preliminary Plan for a New Photographic Search and Surveillance System, 15 Oct 65, quoting USIB Reqmts Stmts of 27 Jul and 31 Jul 64; see also SAFSP Quarterly Program Review, 31 Dec 64 (hereafter cited as QPR with date).

7. 7. For the Itek affair, see MFR, Col P. E. Worthman, 24 Feb 65; MFR, Worthman, 25 Feb 65; MFR, LtCol H. C. Howard, n/d (prob 25 Feb 65); MFR, B. McMillan, 25 Feb 65; and memo, McMillan to C. Vance, D/SOD, 25 Feb 65, no subj, all in NRO files.

8. Msg, B. McMillan, DNRO, to BGen J. L. Martin, Dir/SP, 24 Jun 65; QPR 30 Jun 65; ltr, McMillan to V/Adm W. F. Raborn, Dir CIA, 3 May 65; ltr, Raborn to C. Vance, D/SOD, 25 May 65, no subj.

9. Memo, B. McMillan, DNRO, to D/SOD and Dir/CIA, 13 Jul 65, subj: New Satellite Search/Surveillance System; memo, W. F. Raborn, DCI, to C.R. Vance, D/SOD, 20 Jul 65, no subj.

10. Agreement for Reorganization of the National Reconnaissance Program, signed by C. R. Vance, D/SOD, and W. F. Raborn, DCI, 11 Aug 65.

11. Memo, D. F. Hornig, Spec Asst to the Pres for Sci and Techn, to C. R. Vance, D/SOD, 30 Jul 65, no subj, in DNRO files.

12. Msg, B. McMillan, DNRO, to BGen J.L. Martin, Dir Progm A, 23 Aug 65.

13. Msg, SAFSP to SAFSM, 12 Aug 65.

14. Msg, B. McMillan, DNRO, to BGen J. L. Martin, Dir/SP, 22 Sep 65; msg, McMillan to Martin, 29 Sep 65.

15. Minutes, Meeting of the NRP Executive Committee (hereafter cited as NRP ExCom) on 6 Oct 65.

16. Minutes, NRP ExCom Mtg of 16 Nov 65.

17. DNRO Action Memo No.1, 15 Oct 65 (signed by A. H. Flax, DNRO): Terms of Reference for the Project Management Task Group for the New Photographic Satellite Search and Surveillance System; NRO Actn Memo No.2, 15 Oct 65: Terms of Reference for the Technical Task Group… (as above); System Operational Requirement, Description, and Preliminary Plan for a New Satellite Photographic Search and Surveillance System, 15 Oct 56 (signed by Col D. L. Carter, chm of task grp).

18. Msg, BGen J. L. Martin, Dir /SP, to Dr A. H. Flax, DNRO, 27 Oct 65; msg, Flax to Martin, 5 Nov 65; msg, Martin to Flax, 7 Nov 65; msg, Flax to Martin. 15 Nov 65.

19. Msg, BGen J.L. Martin, Dir/SP, to EK, 22 Nov 65.

20. Msg, BGen J. T. Stewart, Dir/NRO Staff BGen J.L. Martin, Dir/SP, 7 Dec 65; msg, Stewart to Martin, 8 Dec 65; DNRO Actn Memo No 6, 7 Dec 65; memo, Col D. L. Carter, Task Grp Chm, to A. H. Flax, DNRO, 28 Jan 66, subj: RFP for the Photographic Subsystem for a New Search/Surveillance System; QPR, 31 Mar 66.

21. Minutes, NRP ExCom Mtg of 26 Apr 66; memo, BGen J. L. Martin, Jr. Du/SP, to DNRO, 4 Nov 65, subj: Comments on Alternative Management Arrangements for the New Photographic Search and Surveillance System (in SAFSS files).

22. Memo, A.H. Flax, DNRO, to D/Sec Def, DCI, Spec Asst to Pres for Sci and Techn, 22 Apr 66, subj: New General Search and Surveillance Satellite System; memo, Flax to BGen J. L. Martin, Dir/SP, 30 Apr 66, subj: Implementation of Hexagon Program.

23. See memo, Flax to D/Sec Def et al, 22 Apr 66, and incls, DNRO files.

24. QPR, 30 Jun 66.

25. QPR, 30 Jun 66; msg, BGen J. L. Martin, Dir/SP to A.H. Flax, DNRO, 6 May 66; msg, Flax to Martin, 13 May 66; msg, Flax to Martin, 3 May 66.

26. Ltr, H. W. Paige, VPres and GenMgr, GE Miss and Space Co, to A.H. Flax, ASAF (R&D), 14 Jun 67, no subj (reviewing "points made last year"); ltr, Flax to Paige, 29 Jun 67, both in DNRO files.

27. QPR, 30 Jun 66, memo, A.H. Flax, DNRO, to Hexagon System Project Office, 25 May 66, subj: Instructions for Satellite Basic Assembly Source Selection; memo, BGen J. L. Martin, Dir/SP to Flax. 16 Jun 66, subj: Re-entry Vehicle Study for the HEXAGON System; msg, Flax to Martin, 6 Jul 66; memo, Flax to Re-Entry Vehicle Source Selection Advisory Council, 6 Jul 66, subj: Instructions for Re-Entry Vehicle Source Selection Study.

28. QPR 30 Sep 66; memo, L. C. Dirks, Chmn, Sensor Subsystem Source Selection Board to DNRO, 30 Aug 66, subj: HEXAGON Sensor System Source Selection Board Recommendations; msg, DNRO to BGen J .L. Martin, Dir/SP, 11 Oct 66.

29. NRP ExCom Minutes, 17 Aug 66, 23 Nov 66, 16 Dec 66, 17 Nov 67.

30. QPR's of 30 Sep 66 and 31 Dec 66; minutes, NRP ExCom mtg of 23 Nov 66; memo, J. Q. Reber, Secy, NRP ExCom, to NRP ExCom, 9 Dec 66, subj: Agenda for NRP ExCom Meeting of 16 Dec 66.

31. QPR, 31 Mar 67; msg, A.H. Flax, DNRO to BGen J.L. Martin, Dir/SP, 21 Feb 67; QPR, 30 Jun 67; msg, Flax to Martin, 8 May 67.

32. Msg, A.H. Flax, DNRO, to BGen J .L. Martin, Dir/SP, 14 Jul 67; msg, Flax to Martin, 19 Jul 67; QPR, 30 Sep 67, 31 Dec 67, 31 Dec 68; minutes, NRP ExCom, mtgs, 17 Nov 67 and 20 Dec 67.

33. QPR, 31 Mar 68 and 30 Jun 68; msg, A.H. Flax, DNRO BGen J. L. Martin, Dir/SP, 20 May 68; msg, NRO Compt, to LtCol J. McBride, SP, 10 Jun 68.

34. Minutes, NRP ExCom mtg of 20 Aug 68.

35. BoB Position Paper, "The Need for the Hexagon Photographic Satellite," Nov 68, in NRP ExCom files. (Holograph notes by A. H. Flax, DNRO, on margins of file copy.)

36. Ltr, R. P. Mayo, Dir/BoB, to R. Helms, DCI, 22 Mar 69, no subj; ltr, L. A. Bross, CIA, to J. L. McLucas, DNRO, 4 Apr 69; ltr, Mayo to R. M. Nixon. Pres, US, 21 Apr 69, subj: FY 1970 Intelligence Program Savings, with incls. See also memo, BGen R.A. Berg, Dir NRO Staff, to McLucas, 28 Apr 69, subj: BoB Paper on HEXAGON (All in NRO files)

37. *Endnote Redacted*

38. See Ch III, Vol I, this history for an account of the final Corona program extension proposals (1970-1971).

39. Minutes, NRP ExCom Mtg of 15 Jul 71.

40. QPR's, 30 Sep 68, 31 Dec 68, 31 Mar 69; minutes, NRP ExCom mtg of 20 Aug 68 and 13 Nov 68. The engineering review was conducted by a special committee headed by Dr A. F. Donovan of Aerospace Corp: see rpt. 15 Jan 69.

41. Memo, F. R. Naka (D/Dir NRO) to J. L. McLucas, DNRO. 4 Sep 69, subj: Report of HEXAGON Review Committee, with atchd rpt, 20 Jun 69.

42. Minutes, NRP ExCom Mtg of 8 Aug 69.

43. Rpt, Second Report of HEXAGON Review Committee, 4 Nov 69; minutes, NRP ExCom Mtg of 25 Nov 69; Third Report of HEXAGON Review Committee, 22 Jan 70; memo, F. R. Naka. D/DNRO, to DNRO, 28 Jan 70, subj: 2nd and 3rd Reports of the HEXAGON Committee; memo, J. L. McLucas, DNRO, to NRP ExCom, 2 Feb 70, subj: Adequacy of the CORONA/HEXAGON Overlap.

44. QPR, 31 Dec 69.

45. Rpt, Director's Report to the NRP Executive Committee on FY 1970 Status, FY 1971 Program, by J. L. McLucas, DNRO, 15 Jul 70; QPR's 31 Dec 69, 31 Mar 70, 30 Jun 70.

46. Memo, F.R. Naka (D/DNRO) to J. L. McLucas, DNRO, 31 Jul 70, subj: HEXAGON.

47. Minutes, NRP ExCom Mtg of 17 Jul 70.

48. Memo, NRO Staff, to Col E. Sweeney, Dir NRO Staff, 31 Aug 70, subj: HEXAGON.

49. Msg, BGen W.G. King, Dir/SP, to J.L. McLucas, DNRO, 15 Sep 70; memo, McLucas to USIB, 18 Sep 70, subj: HEXAGON Status.

50. QPR, 30 Sep 70, 31 Dec 70; msg, King to McLucas, 15 Sep 70.

51. Minutes, NRP ExCom Mtg of 29 Jan 71.

52. Msg, Col F.S. Buzard. Hexagon program mgr, to BGen L. Allen, Dir/SP, 26 Apr 71; msg, Allen to F.R. Naka, D/DNRO, et al, 27 Apr 71; memo, Hexagon Sensor Sys Progm Dir to D/DNRO. 30 Apr 71, subj: Transmittal of Shutter Replacement Schedule.

53. See Minutes, NRP ExCom Mtg of 13 Jul 71.

54. For details of operational responsibilities and related matters, see Rpt, Hexagon Concept of Operations, prep by SOC and publ by NPIC, Sep 70.

55. San Jose News, 20 Mar 71.

56. See, for example, San Francisco Chronicle, 16 Jun 71; msg, Dir /SP to NRO staff, 25 May 71; ltr, J. L. McLucas, DNRO, to R. Helms, DCI, 14 Jun 71, no subj.

57. Msg, J. L. McLucas, DNRO, to Dir/SP, et al, 22 Jun 71; rpt, Report [of the DNRO] to the President's Foreign Intelligence Advisory Board on the National Reconnaissance Program, July t, 1970 to June 30, 1971, prep by NRO staff, 1 Jul 71.

58. Msg. Col F.S. Buzard, Hexagon SPO, to J. Hughes, CIA, 16 Sep 71; ltr, A. Lundahl, NPIC, to J. L. McLucas, DNRO (holograph ltr), 13 Jul 71 (incorrectly dated 1970 on original), no subj; QPRs, 30 Jun 71, 30 Sep 71; memo, LtGen D. V. Bennett, Dir DIA, to DNRO, 14 Jul 71, subj: Photographic Satellite Launch Schedule.

59. Memo, Bennett to McLucas, 14 Jul 71, QPR, 30 Sep 71, 31 Dec 71.

60. MFR, D. W. Patterson, Hexagon Sensor Subsys Prog Dir, 28 Jul 71, subj: RV-3 Recovery Planning Meeting: memo, to J. L. McLucas, DNRO, 3 Aug 71, subj: Possible Recovery of HEXAGON mission 1201 RV-3.

61. Interview, LtCol F. L. Hofmann, NRO Staff, by R. L. Perry, 7 Jun 73; msg, Col F. S. Buzard, Hexagon Prog Dir, to Col D. D. Bradburn, Dir NRO staff, 7 Jan 72; memo, USN, NRO Staff, to J. L. McLucas, DNRO, 9 May 72, subj: Deep Sea Recovery of HEXAGON; MFR, NRO Staff, 14 Sep 71, subj: Status of Recovery Effort as of 14 September 1971; memo, J. L. McLucas, DNRO, to Asst Sec Navy (R&D), 15 May 72, subj: Deep Sea Recovery of HEXAGON Reentry Vehicle; rpt, Report to the President's Foreign Intelligence Advisory Board on the National Reconnaissance Program, July 1, 1971 to June 30, 1972 (aprox 30 Jul 72).

62. MFR, Sat Ops Center, 8 May 72, subj: Trip Report—Mission 1202 Critique; msg, all EK to Hexagon Progm Office et al, 30 Jan 72, 13 Feb II, 21 Feb 72; minutes, NRP ExCom Mtg, 23 Nov 71; memo, Bennett to DNRO, 14 Jul 71; rpt, Director's Report to the NRP Executive Committee on Current Status and FY 72-77 Fiscal Program, prep by NRO Staff, 11 Nov 71; QPRs, 31 Dec 71, 31 Mar 72.

63. Memo, LtCol H. B. Peake, NRO Staff, to J. L. McLucas, DNRO, 21 Aug 72, subj: Aft Camera Problem; QPR's, 31 Dec 71, 31 Mar 72, 30 Jun 72, 30 Sep 72.

64. QPR's, 30 Sep 72, 31 Dec 72; MFR, LtCol H. B. Peake, NRO Staff, 18 Apr 73, subj: HEXAGON Program Review—16 April 1973.

65. QPR's, 31 Mar 72, 30 Sep 72, 31 Dec 72, 31 Mar 73 and 30 Jun 73; memo, LtCol F.L. Hofmann, NRO Staff to Col J. Shields, 22 Jan 73, no subj.

66. Minutes, NRP ExCom Meetings of 23 Nov 71, 19 Jul 72, 27 Sep 72; memo, R. E. Williamson, Lockheed, to Office Dir/Spec Proj, 16 Nov 71, subj: Dr. Sorrels Briefing on 12 Nov.

67. MFR, no sig, 15 Sep 71, subj: HEXAGON Block II, atchd to note, to NRO Compt, 16 Sep 71, no subj.

68. MFR, 15 Sep 71; note, to NRO Staff, 16 Sep 71, no subj.

69. Informal note, NRO Staff, to Maj, 27 Sep 71, subj: HEXAGON Block II.

70. Msg, BGen L. Allen, Jr, Dir/SP, to J.L. McLucas, DNRO, 22 Nov 71, and CIA SSPO to Dir/SP, 19 May 72.

71. Msg, BGen D. D. Bradburn, Dir/SP, to J. L. McLucas, DNRO, 20 Apr 73; msg, McLucas to Bradburn, 4 May 73; memo, Col J. E. Kulpa, Dir NRO Staff, to McLucas, 1 May 73, subj: HEXAGON Panoramic Camera Improvement Study.

72. Memo, M.R. Laird, SecDef, to Pres U.S., 17 Aug 71, subj: Readout Satellites; memo, H.A. Kissinger, Spec Asst to Pres, to Sec Def, et al, 23 Sep 71, no subj; memo, Col D. D. Bradburn, Dir NRO Staff, to J. L. McLucas, DNRO, 22 Oct 71, subj: Transfer of HEXAGON Sensor Subsystem Contracts from OSP to SAFSP; memo, F. R. Naka, Dep/ DNRO to McLucas, 21 Oct 71, subj: Hexagon Transfer.

73. Rpt, Hexagon Transition Plan, Mar 72, prep by D. L. Haas (CIA), (CLA/SSPO), Col R.H. Krumpe (Prog A), with concurrence of (CIA/ Dir Rec Progms) and approval of BGen L. Allen, Jr (Dir/Prog A); msg, DNRO to Dir/CIA Recce Progms and Dir/Prog A, 29 Nov 71.

74. Msg, L. C. Dirks, CIA, to BGen D. D. Bradburn, Dir /Prog A and J. L. McLucas, DNRO; Bradburn to Dirks and McLucas, 30 Jun 73.

www.ingramcontent.com/pod-product-compliance
Lightning Source LLC
Chambersburg PA
CBHW080550170426
43195CB00016B/2745